2018 Amateur Radio Emergency Communications Symposium Gainesville, Florida

Co-Sponsored by

Alachua County ARES

Santa Fe College Amateur Radio Society

Edited by

Gordon L. Gibby KX4Z NCS521

ISBN-13: 978-1983678806

ISBN-10: 1983678805

DEDICATION

This Report of the 2018 Amateur Radio Emergency Communications Symposium is dedicated to all the hard-working volunteers who made this effort possible.

HF VHF emergency station in a Full Scale Exercise

CONTENTS

ACKNOWLEDGMENTS

I'd like to acknowledge all the people who put together the talks that made up this conference:

Joe Bassett W1WCN
Jeff Capehart W4UFL
Susan Halbert KG4VWI
Dave Welker W2SRP
Vann Chesney AC4QS
Art & Cindy Grant KM4YGH, KM4YGG

Thanks!!

Gordon Gibby KX4Z NCS521

INTRODUCTION

The 2018 Emergency Symposium started out as an idea by Jeff Capehart to hold a "Digital Training" for outlying Emergency Coordinators.   When Jeff first thought of it, our Alachua County ARES group had relatively little outreach to other counties, so the idea didn't go that far.

However, as time went on, we developed increasing contacts with interested volunteers in other counties, and our own members grew in their understanding and capabilities with emergency communications – and I was able to get DHS AUXCOMM training thanks to a group of Alabama volunteers who worked to hold a session.   Hams helping other hams by holding training sessions seems to be the way Emergency Communications is growing the fastest.

So it began to seem more possible to hold just such a training session --- but more broadly than just "digital", to include as much as possible about amateur radio emergency communications.   We're blessed locally to have an active Emergency Operations Center and a wide range of local amateur radio talent.

So an email was drafted to send to all known North Florida Emergency Coordinators and their assistants and this conference began to take shape when several positive responses were received.

The goal of this conference is to move participants into the ability to hold Full Scale Exercises --- that is truly where the "rubber meets the road" and hams begin to assess their capabilities for real.   Your group can start small, but if you will take the time to simply write down an ICS201 Incident Briefing scenario, conduct a TableTop exercise first so your participants have a chance to try out the techniques you'd like them to exercise --- your group will have a blast with your Full Scale Exercise and move ham radio miles down the road in your county toward better preparedness to serve your community (or another community!) in a time of need.

Microwave digital station during a Full Scale Exercise

1   WHAT QUALITIES MAKE GOOD LEADERSHIP? – BRAINSTORMING

by Vann Chesney (AC4QS)

Integrity/Honesty
The most fundamental quality of a leader is honesty and integrity because this is the way to gain the trust of followers. There is no leadership without trust.

Courage/Confidence
A leader has to have the courage to step up when needed. Having confidence in yourself inspires confidence in others.

Vision/Focus
A leader has to have the vision to determine what is most important. To bring focus to chaos.

Decisive
To be an effective leader you must be able to make appropriate decisions in a timely manner.

Inspire
An essential quality for any leader is the ability to inspire others.

Topics for Discussion

Local Leadership
What are examples of leadership in your local areas?

_____
_____
_____
_____

Incident Leadership
What are examples of leadership in actual emergencies?

_____
_____
_____
_____

NOTES

2 EMERGENCY COMM SCENARIOS:
THE "WHAT-IF?" APPROACH

The very first sentence in the FCC Part 97 regulations states the purposes of the Amateur Radio Service, and first among those is:

> (a) Recognition and enhancement of the value of the amateur service to the public as a voluntary noncommercial communication service, **particularly with respect to providing emergency communications** (emphasis added)

If emergency communications are one of the most important reasons that we are granted access to such huge and valuable chunks of radio spectrum, in just about every portion of the radio spectrum, *then it would be not only altruistic, but WISE for amateur radio operators to give careful heed to actually being prepared for emergency communications.*

There is certainly a valid place for what I call "Fair-Weather Fun" public service communications-- parades, marathons, etc. These give a wonderful public service, a chance to practice elementary skills, develop organization, among other good effects.

But at some point, we need to be thinking about the "All H-E-L-L Just Broke Loose!" emergency communications, and be ready for those, as well. Armies are in place for the defense of the nation, and they don't just practice for assisting in food drives and toy collection --- they practice for WAR. Hams should be practicing for real-world emergency communications!

One way to spur your group toward progress in this area is to conduct a brainstorm session based on the "What-If?" approach.   In my field (medicine) we plan for backup plans if a procedure doesn't go just the way expected --- we still need to care for the patient. That requires a bit of thinking about what kinds of things can go wrong --- and those are exactly the kinds of questions we pose to our resident physician trainees.

In the field of emergency communications, here are some starting points for scenarios:

1. **Weather**: Hurricanes, multiple tornadoes, ice storms (bend microwave gear and towers to uselessness); solar weather including coronal mass ejection (CME)
2. **Utilities**:   Loss of power generation station, loss of portions of the electrical grid; malfunction of piped natural gas
3. **Accidents:** Wildfires, nuclear spills, nuclear meltdown, hazardous waste dispersal
4. **Attack**:   Power or communications grid hacking; loss of data interconnections; nuclear attack; electromagnetic pulse (EMP) attack; biologic attack (anthrax, novel viruses, etc); chemical attack (neuro-paralytic agents, etc)

In all of these there are foreseeable loss of communications which leads to **chaos** and loss of "command and control" by the authorities charged with providing assistance. It was one of the most devastating problems faced during Hurricane Katrina in New Orleans. During Katrina, despite days of advance warning,

- 37 of 41 broadcast radio stations in and around New Orleans were knocked off the air.
- Thirty-eight 911 call centers went down *So much for capturing citizen cries for help!*
- 20 million phone calls did not go through on the day after the hurricane.
- More than 3 million telephone lines went down. *Can't even call for help -- or direct aid workers*
- The police headquarters of New Orleans, as well as 6 of 8 police district buildings, were out of commission.
- New Orleans police communications system failed and remained failed for three days. Only 2 channels of a backup system remained for **hundreds** of emergency personnel who were already overwhelmed.
- Louisiana state police communications were dependent on 46 towers to support 10,000 users. After loss of electrical distribution, tower after tower ran out of power when their stored generator fuel was exhausted. Refueling was nearly impossible due to flood waters and debris.(Communications, <u>A Failure of Initiative. Final Report of the Select Bipartisan Committee to Investigate the Preparation and Response to Hurricane Katrina</u>, February 15, 2006. Accessed at: <u>https://www.gpo.gov/fdsys/pkg/CRPT-109hrpt377/pdf/CRPT-109hrpt377.pdf</u>)

Police and public safety officers lost trunking systems and were limited to only a couple simplex frequencies. Units were found who had not had communications with a superior officer in days! Mississippi National Guard were reduced to using *runners*! (Major General Harold A. Cross, Adjutant General, Mississippi National Guard. <u>A Failure of Initiative. Final Report of the Select Bipartisan Committee to Investigate the Preparation and Response to Hurricane Katrina</u>, February 15, 2006. p. 174. Accessed at: <u>https://www.gpo.gov/fdsys/pkg/CRPT-109hrpt377/pdf/CRPT-109hrpt377.pdf</u>) How do you manage a complex emergency response involving tens of thousands of workers, and possibly millions of assets? Databases and logistics are near to useless when there is no INFORMATION at all. (<u>http://www.qsl.net/nf4rc/KatrinaComms.pdf</u>)

During a disaster, communications need to flow in basically the same directions they do normally: lots of short-range communications within the city or disaster area, as people try to reach authorities, and authorities try to reach providers (police, fire, ambulance). Broadcast communications to give wide-area instructions and information to the population to reduce or quell rumors. Long distance communications between the disaster area and support, response, resupply areas in unaffected areas. Amateur radio operators are familiar with both short and long distance communications, and both were needed in the Puerto Rico early response.

During a disaster, information flows include a plethora of short "tactical" communications to pass along simple tidbits of position or situation information and response or commands, and also include larger data-type information needed to handle logistics of an ever-increasing disaster response --- the sort of stuff that drives databases and tracks thousands to millions of assets responding. Amateur radio operators are typically more familiar with the short "tactical" communications that fit well with VHF FM voice and repeaters – indeed, this is exactly the type technology that fire, police and other public service typically use day to day in non-disaster conditions.

Amateur radio operators are often less familiar with long-haul detailed data communications that might be tens to hundreds to millions of bytes of information that must be communicated flawlessly. High bandwidth digital communications are normally accomplished using satellite technology today (think: the gas station pump that almost instantly checks, accepts and bills your credit card) --- and amateurs typically don't have that huge infrastructure in place. Cell phone towers took a decade to dot the landscape and before that satellites took decades to develop and before that microwave towers....   Hams often just don't have that kind of high data rate infrastructure.

In the early portion of a disaster that limitation is less important because the really important communications are disaster related and not the gigabytes that are flowing all around us to keep a society functioning. <u>But Puerto Rico underlined the need to return the high data rate services as fast as possible, because disaster services don't come close to providing the entire range of day-to-day services to a population that a functioning economy provides</u>. Thus military satellite dishes quickly pop up and major-player telecommunications corporations put in emergency dishes, emergency towers and begin to return normalcy.   Citizens today almost all have portable high data rate digital multi-purpose radios on their person at all times --- cell phones!   And when those don't work, society is crippled. Businesses can't function, hospitals are hampered, fuel doesn't flow, bills don't get paid, docks don't get unloaded --- you get the idea why high data rate communications have to be restored as fast as possible. The planned FirstNet system may be just as vulnerable as the cell towers. (Congressional Research Service. The First Responder Network (FirstNet) and Next-Generation Communications for Public Safety: Issues for Congress. https://fas.org/sgp/crs/homesec/R42543.pdf)

One area yet relatively unexplored in amateur radio emergency communications is therefore high data rate WIFI communications on ham or pubic 2, 3, 5 GHz bands using the inexpensive and widely available Ubiquiti line of WIFI transceivers.   It may be the wave of the future to put up mountaintop-to-mountaintop WIFI service quickly to constitute an ad hoc data network for an island or portion of a state. (https://www.aredn.org/ ; our group exercised this type technology in a recent Full Scale Exercise: http://www.qsl.net/nf4rc/CreateSpaceEntireSteinhatcheePlanningAndReport.pdf)

Ham radio operators need to take all of the above into account in their planning for disaster communications and develop not only voice tactical communications, but also

- message relay skills
- digital traffic skills such as FLDIGI / WINLINK (https://winlink.org/)
- HF skills on multiple modes
- wide bandwidth skills on WIFI equipment including tcp/ip networking.

There is quite a lot of ground to cover --- enough to fill the educational calendar of any ARES group!

Operating position in a school shelter during Hurricane.

NOTES

_____
_____
_____
_____
_____
_____
_____
_____
_____
_____
_____
_____
_____
_____
_____
_____
_____
_____

3   LESSONS FROM PUERTO RICO

Reprinted with permission from
"The Day That We Saved Amateur Radio:
a look at amateur radio's role following hurricane Maria in Puerto Rico."
Joe Bassett

"Those who wish to be successful must ask the right preliminary questions." Aristotle Metaphysics II

Amateur radio emergency communication is undergoing a fundamental shift. For decades, success in amateur radio emergency communication scenario has been measured in the number of messages transmitted, received, and accounted for. In that context, the primary information transmitted via high frequency radio waves is Health-and-Welfare or, as defined by the American Red Cross, Safe-and-Well messages. By definition, these messages are innocuous in nature and require no action beyond acknowlededgment. These messages certainly bring comfort and consolation to concerned family members outside the areas affected by a particular crisis. However, the prevalence of social media, such as Facebook's "I'm safe" button, or emerging technology, such as the VSAT devices employed by the American Red Cross, raise the question, "Is high frequency radio communication, vis a vis amateur radio, still the primary, or most effective, means of conveying this type of message?"

Of course, the underlying premise of ham radio's mantra "when all else fails" still applies. Not only is there the possibility of a catastrophic event rendering most forms of communication inoperable, the effects of hurricanes Irma and Maria on the island of Puerto Rico rendered it a reality. Unfortunately, the same 135 miles per hour winds, that laid waste to Puerto Rico's infrastructure, also made transmitting "Safe-and-Well" messages premature. In essence, in the immediate week's after hurricane Maria's landfall a large portion of Puerto Rico's population wasn't safe and well. The people of Puerto Rico had immediate need for life saving measures such as shelter, water, food, medical care, portable power solutions, etc. Safe-and-Well messages could be sent *after* people were safe and well.

Yes, in the immediate days following landfall, there was a trickle of Health-and-Welfare information transmitted by intrepid ham radio operators in Puerto Rico to the U.S. mainland via the Salvation Army Team Emergency Network (SATERN) and some *ad hoc* operators on 14.270Mhz, but the need for immediate, tactical communication for recovery efforts was paramount. The necessity for, and value of, tactical communication is underscored by the selfless efforts of Puerto Rican hams that stood up a VHF net in support of the Puerto Rico Electric Power Authority within hours of Maria's exit off the northwest coast of the island. In the face of personal hardship these operators, by all accounts, provided an invaluable service to early recovery efforts.

In due course, Safe-and-Well messages became the order of the day and proved invaluable in aid and comfort, but by the time transmission of these messages attained wide-spread necessity, means of communication by means other than high frequency radio waves were possible and even better suited for the task. This is not to say that transmission of Safe-and-Well messages via radio wave wasn't accomplished, and beneficial, in the early weeks of recovery, but it wasn't the pressing communication need.

Against the backdrop of this immediate need for tactical communication, throughout an island measuring more than 3,500 square miles, the American Radio Relay League (ARRL) team (Force of Fifty) arrived in San Juan under the auspices of the American Red Cross (ARC). As we'll see the Force of Fifty was equipped with a template, tasked with a single plan and equipped exclusively for *that* single plan. In defense of the ARRL, and the American Red Cross, the extent of devastation wrought on Puerto Rico's communication infrastructure was unprecedented. The nature of the disaster made it necessary to anticipate need far in advance, in some cases days and weeks in advance. Under the best circumstances, it's nearly impossible to anticipate, with any certainty, the communication needs following a large-scale disaster. Even so, advance teams can advise inbound personnel as to immediate needs in the first hours following impact. The utter absence of inter-island communication rendered actionable data-collection, and the relay of material requests to the mainland, left strategic response tenable, at best.

With that, the ARRL and ARC made their "best guess" for communication needs in Puerto Rico. The initial mission for the Force of Fifty was to embed with Red Cross teams, interview Puerto Rico residents, populate a questionnaire of Safe-and-Well information, utilize high frequency radio to transmit that information to an online database stateside. All in all, it was an elegant solution utilizing portable high frequency radios, inexpensive and workable digital capabilities (Winmor Winlink), and 40-meter dipole antennas with automatic "tuners." Unfortunately, this solution was limited in scope for the stated task, more on this later. Furthermore, this strategy was singularly ill-suited for tactical communication. In essence, these radio "rigs" were uni-taskers in a situation that required multi-taskers.

Keeping in mind that the original ARRL/ARC mission relied on high frequency communication with stateside RMS (Winlink) stations, propagation over at least 1000 miles was necessary. This, of course, assumes that the closest station was always available. There are several stations, occupying multiple frequencies, located within 2000 miles of Puerto Rico. Many of these stations proved accessible, and were utilized, for Winlink communication. However, reliability was limited, exacerbated by the use of Winmor rather than Pactor.

The necessity of erecting antennas less than one-half wavelength above ground proved problematic for 40-meter propagation, outside of near vertical incidence skywave (NVIS), usually 30-400 miles. Even utilized as NVIS antennas the dipoles presented the challenge of overcoming increased standing wave ratio (SWR).

This reveals another facet of the "outdated" paradigm: rigidity or trying to force the situation to meet the tools and skillset. The obstacles outlined above could have been, and in some instances, were, foreseen. It is common ham radio operator knowledge that 40-meter daytime propagation over more than 500 miles is undependable. The lack of structures capable of supporting a flat-topped dipole at one-half wave length exacerbates the issue. It is stipulated that using the auto-tuners provides an impedance match at the radio, however this still results in a non-resonant antenna at other wave lengths. None the less, the tools deployed to Puerto Rico were forced into the mission. The limitation of 40-meter propagation for 1000 mile plus transmissions of digital signals at 150-300 baud should have, and in some quarters, was, foreseen. Regardless, it was employed as a tactic.

Ironically, this same limitation reveals evidence of a fundamental change in thinking; a thinking that engenders the "can-do spirit" of amateur radio: how can the tools at hand serve the mission? These antennas were limited in supporting long-range transmissions, however, in the absence of VHF equipment, the 40-meter dipoles could be repurposed for tactical communication as NVIS antennas providing adequate communication within the area of Puerto Rico, 150 miles east to west.

Unfortunately, many of the operators deployed with the Force of Fifty lacked adequate knowledge of NVIS properties. While many of the operators were adept at specific skills, few of them were well rounded in amateur radio emergency communication. Some had excellent understanding of constructing antennas in general, and NVIS antennas specifically, but lacked knowledge of Winlink, in at least two cases team members didn't even have Winlink accounts prior to deployment, in spite of Winlink experience being a primary skill listed in vetting documents.

Having said that, amateur radio emergency service is best served by jettisoning an old adage, "Jack of all trades, master of none." Many amateur radio operators are myopic toward the ever-changing landscape of ham radio. As such, they concentrate on mastering one aspect of the hobby at the expense of others, fearing that their preferred component will suffer. On the contrary, the amateur radio emergency communication operator is best served by being "a jack of all trades, but a master of one."

Another way of looking at this change of thought is to compare two premises. One states, "we're radio operators who happen to communicate;" the other, "we're communicators who happen to operate radios." This book will explore amateur radio's contribution to relief efforts in the aftermath of hurricane Maria in Puerto Rico. This exploration, along with some tangential evidence from the author's experience from the last two hurricanes to make landfall in Florida (Matthew and Irma), lead to the fact that effective amateur radio emergency communication is not measured by the quantifiable, "How many messages were transmitted?" Its effectiveness is invaluable, but immeasurable, in the nebulous answer to the question, "How quickly are we able to respond to the myriad of catastrophic possibilities facing our communities?"

SHAMELESS PLUG FROM THE EDITOR

Joe Bassett graciously contributed this chapter from his upcoming book to this Symposium document. He's hoping to get the entire book finished & out soon --- there have been some unexpected obstacles, of course. Might I suggest that it would be a great idea to get his book when it comes out and pass it around your local emergency group, or urge others to purchase it also? The more we draw from others' expertise, the more mistakes we can avoid.... G. Gibby.

4 GETTING GEAR TO THE NEED

VHF, Go Boxes, Power and Power Connections
by Art Grant

Here we are, no dangerous winds, fires, floods or ice. We are preparing for those events like true Volunteers. Have you prepared enough or is there one or more areas that could be a little better? Have you taken an inventory lately of what works and what you do not want to carry around in an emergency. Who do you depend on that can work with you and your equipment?

What is really needed? Power of course! Dependable power and that calls for fuel for enough operation time times 3 to be sure you can operate the entire time. My Honda 2000 watt generator uses 1 gallon every eight of use or three gallons a day times 3 equals at least 10 gallons. Oh do not forget the Stable otherwise your fuel becomes shlack and no go on the generator. Now where to put it, in the garage so no one can remove it to take home----NO. Carbon monoxide will kill you! Put it outside with a good strong chain to secure it with padlocks. You did get those at the store after the last emergency? How about a funnel for refueling or better yet a hand pump for transferring that precious liquid when the unit cools down.

We can now look at how many watts do we need for our radio equipment. Several different sources have many theories. I thought 2000 watts would produce about 15 amps of 120 volts in case I wanted to run a small refrigerator or some lights. Lights, of course LEDs. Be careful with fans and other items like phone rechargers, computer operations besides our radios and amps. We will need to charge our batteries directly then attach the radios and equipment to the batteries. The batteries provide a steady source of constant power to prevent damaging our equipment.

We need deep cycle batteries for both voice and digital. Bigger for voice, smaller for digital as we can transmit more info digitally than by voice taking less electricity. Voice though is our constant, currently but digital is making great in roads. In our group we try so many different types so we can decide on just one or two for everyone to perfect. Even the biggest or best is no good if the average person cannot use them. Find one or two types and teach those first. Some will exceed the expectations but we have to look at the whole team and what they can do as a group not who is the best!

Solar, that power that is always there even after an emergency. The panels have to big enough to charge a deep cycle battery for about 6 hours a day, delivering enough amp hours required for your equipment. Well how much do I need? Answer- How much radio do you have and what additional items do you require? Everyone will have different answers! CT Solar has an information page online at their website under Solar Power for Amateur Radio FAQ. It is a good guideline for true power without issues like clouds, bad connections or wiring. Figure

your needs then purchase a true wave generator not a square wave. That will be addressed later.
I found a 100 watt ready to run PV (solar)system, wired with a controller and power pole connectors, clamps and ready for an add on set of panels for more backup. Price on Amazon from Acopower was $235.00 delivered. Legs and sturdy carrying case included. Again charge the batteries, do not hook directly to radio equipment!

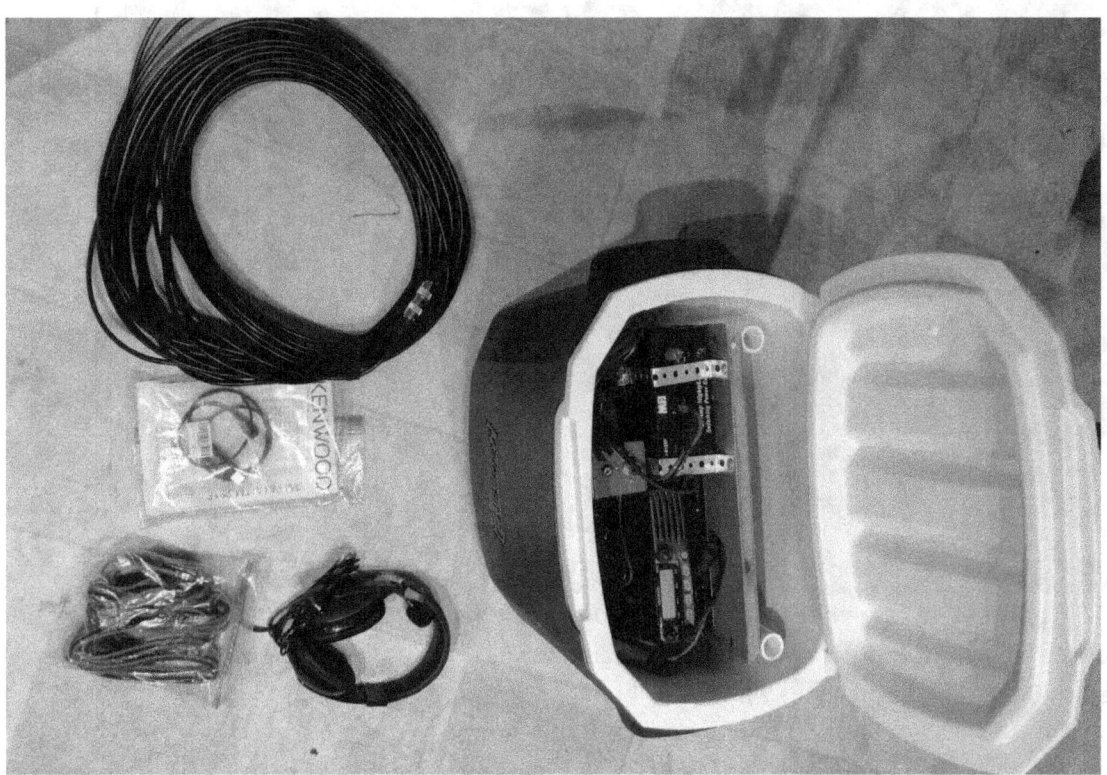

My go box is built with a Coleman Cooler that has wheels and a pull handle so I do not have to tote it in and out of shelters. The radio is removable in order for it to sit on a table and use the cooler as a seat. My wife has another very similar so we can trade equipment if we have a breakdown. How do we accomplish the interconnections. With Power Pole connectors on each piece. Do you have the same connectors in your groups? How can you replace an inop radio without totally replacing the unit. We have Power Gates set up to connect the battery and power supply and power the radio without interruptions during a loss of power. When the radio does not require power for transmitting then the charger goes directly to the back up battery and can reduce to a trickle charge when battery is full.

The EOC's go boxes were assembled hours before Hurricane Irma approached Florida. Several members of our group went to a home of another where the radios and power supplies, power gates were delivered and in 2-3 hours we had 3 operational go boxes with head phones and antenna connectors with SWR meters. The boxes used were water tight storage boxes found at a local home center.

Our antennas were stick antennas that were put up with slingshots and fishing line. We assembled them during one of our training afternoons. Over a year we have dedicated many hours to training and assembling. We got to know people we never would have met and helped each other understand the emergency radio system in our local area. At the beginning I made a stick antenna and today it is mounted in a pine tree 67 feet above ground. It has been used as a node for 2 meter digital for over a year now. Inside we have a raspberry pi computer, monitor and keyboard and mouse.

In troubleshooting an interference problem at our house I studied and built a yagi antenna with guidance from other members. We searched for two months and found in the end, the light in the garage was producing a signal covering up our 2 meter node.

So where does this leave us? Take the challenge of training and building then plan a simple exercise using each of these items we have discussed. During an emergency, there is no training. Just a set of by the seat of our pants, fixes that may cost lives. Have fun doing it, BBQs bring out people and cover dishes are the cement we will remember as we communicate with each other during a true emergency. Share with everyone in groups, homes, Red Cross offices, EOCs and over coffee. We will not have these team building opportunities during a true emergency. I hope I never have to use these new found skills but if I do I will be ready. Will you?

Thanks, Art Grant

HF GO-BOX
Gordon Gibby

The issues of power connections, fusing, etc are similar for any amateur radio station, whether HF or VHF. While there are quite a few QRP HF transceivers on the market, for a portable emergency comms HF station you might be wise to select something that has 100 watts nominal output – and then back it off as conditions and power supply dictates.

Likewise, digital signals for HF aren't that different from VHF, with the exception that PACTOR modems have an application on HF. However due to their extreme costs (> $1000) most volunteers may settle for just a soundcard system such as a Signalink, any of a number of competitor units, or even a homebrew design such as our ARES group has effectively utilized.   See: http://www.qsl.net/kx4z/InexpensiveTNC.pdf   ; http://www.qsl.net/nf4rc/UnderstandingAudioChannelConfiguration.pdf

HF has more significant problems getting a working emergency ANTENNA than does VHF with their ubiquitous whip antennas.   Further, one generally needs to use multiple different high frequency bands (80, 40, 30, 20 meters for example) to accommodate daily changes in propagation.   The choices in portable emergency antennas for HF may be different if all the trees have been knocked down, or depending on whether or not high structures (rooftops) are available.   Generally, one needs either to provision with some form of multiband antenna that presents an acceptable SWR under all conditions, or provide some sort of manual or automatic antenna tuner.

Here's a list of several possible solutions:

- Automatic tuner such as LDG or MFJ equipment: http://www.mfjenterprises.com/Product.php?productid=MFJ-993B http://www.ldgelectronics.com/c/252/products/17/53/1
- Multiband off-center fed dipole with acceptable SWR on multiple bands: https://www.w8ji.com/windom_off_center_fed.htm   https://packetradio.com/catalog/index.php?main_page=index&cPath=49
- Multiband end-fed antenna: http://myantennas.com/wp/product/efhw-8010/

- Wide bandwidth terminated folded dipole (achieves bandwidth by adding loss, but is a reasonable compromise for low-skilled operators): http://www.packetradio.com/t2fd.htm ; http://hflink.com/antenna/#T2FD http://www.tennadyne.com/specs&prices.htm (includes prices)

My personal preference is for an automated tuner such as the MFJ 993B unit, a random length dipole longer than ½ wavelength at the lowest frequency of interest, fed with 300 ohm balanced window line, with one or more high quality 1:1 current baluns (which can be purchased or made easily) to reduce unwanted rf currents in the rig that will upset digital communications. And then lots of ferrite beads on all cables going in the chain from radio to computer, if used. For homebrew baluns see: http://www.qsl.net/nf4rc/BalunPart1.pdf and http://qsl.net/nf4rc/BalunPart2.pdf and http://www.qsl.net/nf4rc/BalunHowTo.pdf

Figure – one example of homebrew balun

Figure: the insides of a popular indoor 1:1 current balun – a toroid with parallel wire

5 AD HOC VHF ANTENNAS

Figure. Very simple VHF antenna made from 2-conductor lampwire (#18, Home Depot).   Rope as insulators on both sides, lampwire as transmission line.   34" dipole showed lowest SWR 1.2 around 143 MHz. Adjust as desired, tape center to set dimension.

by Gordon Gibby

You can be well-served with plenty of options for vhf/uhf antennas using just the following simple designs:

- quarter-wave vertical antenna with a car body / metal roof or something else as the groundplane.
- Horizontal or vertical half-wavelength dipole
- End-fed slim-jim or J-pole (there isn't much difference)

VHF/ UHF antennas are so modestly sized that there isn't much demand for complicated matching systems such as are used on HF to make small antennas "tune" like they are bigger antennas or vice-versa.   The old standard equations for ¼ wave and ½ wave antennas (of uninsulated wire--- insulated wire might require a 5% shorter length or so) from your license exam studying still apply:

Frequency	¼ wave antenna $234/F_{MHz}$ (in feet)	½ wave antenna $468/F_{MHz}$ (in feet)
146 MHz	19 inches	38 inches

220 MHz	12.75 inches	25.5 inches
440 MHz	6 inches	12 inches

To make a ¼ wave antenna, connect the center wire of the coax to a vertical wire of about the length in the table above, and then either make several radials of the same length going out and down a bit, or else use some capacitive coupling from a 4" square piece of aluminum foil or steel or copper closely approximated to your car body, roof or other piece of large metal.

To make a ½ wave center fed dipole (either vertical or horizontal) just make a center insulator out of anything insulating (a piece of plastic, even from a pop bottle, will work) and run stiff wire out each side, equal lengths, so that the total length of the WIRE is as shown in the table above (e.g., 38" for a 2 meter antenna).

If you have an SWR meter (or better, an antenna analyzer) you can then check it and see if the SWR is below 2:1 or so where you need to operate.   If it is better at lower frequences, the antenna is a bit too long; if it is better at higher frequencies, it is a bit too short.   Make small changes (like ½ inch)!

Believe it or not, common lamp cord for lengths up to maybe 25 feet at two meters isn't a terrible feedline and it also makes the antenna as well!   Pull the ends apart, 19" on each side (38 inches total) and tape the middle so it won't separate further, secure everything and connect the "transmitter end" to your transmitter --- and it is likely to work.   Due to the insulation, you may find 36" or so might work better.

VHF/UHF antennas may have broader usable bandwidths than the 3% of center frequency that is common for HF wire antennas, because the thickness of the antenna (wire or tubing) is a greater percentage of the length.

The Slim Jim antenna is a fascinating end-fed antenna with a built-in matching stub, that turns out to be a bit "hardened" against EMP because the stub frankly shorts out lower frequencies.   The matching stub is not really a TRANSFORMER, it is more of a transmission-line based L- C matching system.   The latter point is beyond the scope of this book and not important --- but you will need an SWR meter or better yet an antenna tuner to tune these antennas.   Construction information is as follows:

(Material reprinted with permission from: Amateur Radio Digital & Voice Emergency Communications, by Gordon L. Gibby)

WOODEN SLIM JIM 2-METER ANTENNA CONSTRUCTION DETALS:

Materials:
- pressure treated 1x2 wood from lumber store
- 14-gauge house wire (THHN style, solid conductor) just over 10 feet long.
- Coax line to connect to the matching network.

WIRE: Start with a piece of solid #14 AWG household wire approximately 3 yards and 9 inches long (117")
(It is easier to be a couple inches too long and later nip the excess off.)   Strip the insulation off of 36" at one end..
It is easiest to do this with a pocketknife while holding the wire against a solid flat surface.

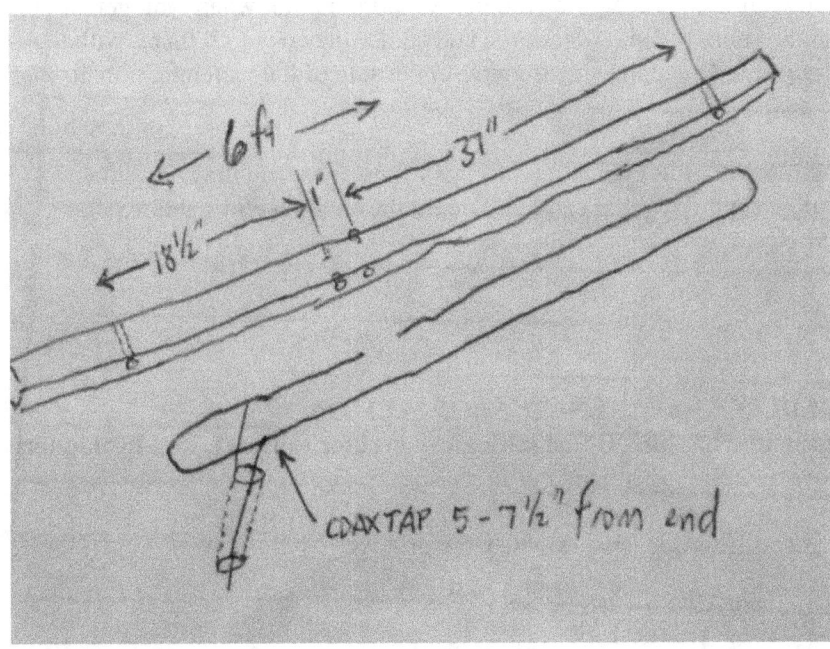

*Fig. Drawing of the wood and wire that make up the antenna.   The end-fed folded dipole is the longer, righthand
portion, while the transmission line matching network is the left hand portion of the wire to which the coax is
attached.* **Note: Making the matching section 19.5" instead of 18.5" sometimes makes this easier to tune.**

WOOD: Start with a pressure-treated 1x2 that is 8 feet long.   These are typically less than $2 at home
improvement stores.   Leave several inches of space (perhaps 8") at one end to "hang" the antenna by, and drill a
1/4" inch hole through from front to back for later hanging.   At 8" from the end (the "top" of the antenna) drill a
1/8" hole clear through from side to side.   37" from that first hole, drill a 1/8" hole just half an inch in to give you
a stopping point for the folded dipole.   Another 19.5"(up to 20.5") further down the wood, drill another 1/8" hole
clear through for the shorting leg of the matching transmission line.   The total length between the two through-
and-through holes will then be 18.5" (to 19.5") (matching section) + 1" (gap) + 37" (folded end fed dipole) =
56.5 (or 57.5) inches, just a bit under under 6 feet).

Distances just aren't that critical.   That 37" inch length is simply not critical.   I tested 3" longer and 3" shorter
and they still would work with just different coaxial cable attachment points, very little difference "shorter" and
so with somewhat more difference with "longer".   The matching transmission line distance probably isn't
terribly critical, either, and increasing it to 19.5" has made better SWR's more easily attainable on several
antennas.

CONSTRUCTION: Thread the wire through as needed to form the antenna as shown in the drawing, and
secure it with electrical tape.   Try to pull sections reasonably tight so the wire hugs the sides of the 1x2 wood.
Cut off any excess as needed so that there is roughly a 1" gap between the free end of the matching line and the
far end of the folded dipole end.

MATCHING: Using an antenna analyzer (with a very short connection, like 2-3" of wire) or an SWR meter (if possible, with a short connection, or a connection that is ½ wavelength (roughly 31" for RG8X) so the impedance isn't altered by your coax line), run the connection up and down the matching section. Use a finger on each side to make the connection, and keep them even with each other. You'll quickly find the point where you get an SWR very near to 1:1, often about 7" from the shorted end. Mark this, and solder the coax to it there, with the center conductor of the coax going to the longer side and the shield going to the side of the matching transformer that goes nowhere.

WATERPROOF: Use a liquid or grease sealant of your choice on the ends of the coax, run the coax either directly away from the matching loop or tape it right down the center. Secure the antenna wires and matching section every 12 inches or so with electrical tape.

VHF/ UHF FEEDLINE REFERENCE INFORMATION

Note that losses increase dramatically if the line has an SWR significantly greater than 1:1, and for ladder line, if the line is wet or covered with snow or ice.

TABLE 7-2. Transmission Line Loss Characteristics

Type of Transmission Line	100 foot loss in dB at 146MHz
RG58A/U (50 ohms)	6.1 dB
RG8X (50 ohms)	4.5 dB
RG 8 /LMR 400 (50 ohms)	1.5-2 dB (LMR is lower loss than RG8)
450 Ohm ladder line	0.4 dB

Remember! These loss values are when the transmission line is operated with a perfect match-- 1:1 SWR. When the SWR is higher, there are points with much higher voltages and much higher currents, resulting in significantly greater losses, particularly for COAX lines. Open wire feeders do much better with high SWR's, but are much more susceptible to losses due to rain, snow, or ice.

6 AD-HOC HF ANTENNAS & BALUNS

by Gordon Gibby KX4Z

Teams were sent to Puerto Rico with one 40-meter antenna and an HF transceiver that had some modest tuning abilities. 40 meters is somewhat a "do-everything" band, but reports indicated that other frequencies were also utilized and teams found ways to make their antennas more versatile. Putting together an emergency HF antenna for any required frequency is an important skill for emergency-oriented amateur radio volunteers.

MAKE A DIPOLE OF ANY KIND

The basic resonant dipole is the foundational antenna, with a length (for uninsulated wire) of about 468/f, where f is in megahertz and the resulting distance is in feet, and the feedline will be attached to an insulator in the middle. If you use insulated wire (like house-wiring), the required length will be a few percent shorter. An SWR meter will help you correct the length; SWR's below 2:1 are preferable, but below 3 works. Insulators can be made out of anything nonconductive --- wood, PVC pipe, plastic, even glass.

Dipoles can be situated horizontally, vertically, or one part vertical and another horizontal, inverted V, or even upright vee --- and they will still work. Horizontal dipoles send their energy out at higher angles of elevation, vertical dipoles at lower angles of elevation. The additive/subtractive effects of the signal that impacts the ground, reflects back up to join the original wave changes the angle of elevation peak; antennas a wavelength or high have lower angles of elevation; antennas more like 1/8 – ¼ wavelength above ground have higher angles, which makes them more useful for NVIS.

Low Dipoles	Higher angle of elevation of maximum energy, work better for hundreds of miles
Really high Dipoles	Lower angle of elevation of maximum energy, work better for thousands of miles
Vertical dipoles	Very long angle of elevation, work better for many thousands of miles

FANCIER ANTENNAS

Simple resonant dipoles tend to have an input impedance in the 50-75 ohm range (depends on height, etc) thus matching coaxial cables well. They will have an acceptable SWR over about about a bandwidth of 2% of their center resonant frequency. Changing to an off-center dipole, by moving the feedline to 33% of the length and the input impedance rises and the antenna also become usable on more different bands. Many hams do well with a 4:1 balun at the feedpoint and coax from there to the transceiver. There are several commercially constructed Windom type antennas using this setup. For example: https://www.amateurradiosupplies.com/windom-antennas-s/55.htm

Random length: Make the antenna a random length, feed it with balanced feedline and use some sort of antenna tuner (manual or automatic) and you have a very low-loss antenna that with experience can be utilized on

23

a wide range of frequencies.   Typically you prefer to have the length of a non resonant antenna $> \frac{1}{2}$ wavelength at the lowest frequency of interest.   Balanced "window" feedline can be obtained from many sources.   I prefer to use stranded 300-ohm feedline because it is easier to work with in portable situations: #562 from thewireman: http://www.thewireman.com/antennap.html#562   or   from   DXEngineering: https://www.dxengineering.com/search/product-line/dx-engineering-300-ohm-ladder-line

WHICH BAND TO USE?
HF amateurs should be well versed on the critical frequency and maximum usable frequency and how they vary with the hour of the day.   If not, then just listen on various bands until you hear callsigns in the general region you need to contact!

Critical Frequency	The highest frequency that can be reflected from straight vertically UP, to straight vertically DOWN back to the sending station. This is the highest frequency that will work to reach people in your same town without using ground wave, for example. At night-time in northern hemisphere in winter and low sunspots, it may be as low as 2 MHz.
	The Critical Frequency is measured every few minutes by radio-ionosondes all over the world and reported over the internet. This is the cause of the "chirp" you may occasionally hear moving rapidly through ham bands.
Maximum Usable Frequency	The highest frequency that can be refracted at all – typically just a glancing blow to the ionosphere, coming back to earth many thousands of miles. DX'ers prefer to use a frequency just below the MUF.

It is an important point to know, that as you move to higher frequencies where the length of your wire is more than 2 wavelengths long, the impedance tends to even out, or stabilize, and matching to it becomes much easier with modern matching systems.

The Mysterious BALUN
Baluns are a way to avoid unwanted radio frequency currents running over what you thought were current-free ground connections, and are generally not that necessary if all you are doing is voice or CW communications. Once a computer is involved, these unexpected currents have a way of causing semiconductor junctions in interface systems (such as USB ports) to be activated at random moments and generally freeze programs, lock transmitters into "transmit" and crash computers.   People who do a lot of digital communication become much better at understanding how to use baluns (and their close relative, the ferrite bead) to reduce these unwanted currents by adding inductance in their unwanted paths.

What is inside a popular coax-to-coax balun (or "un-un")

A homemade current balun that does the same thing.

You can read more about baluns in several places, including:
http://www.qsl.net/nf4rc/BalunPart1.pdf
http://qsl.net/nf4rc/BalunPart2.pdf
http://www.qsl.net/nf4rc/BalunHowTo.pdf

If you do digital HF communications and don't wish to homebrew your own baluns, you probably should purchase one or more of the following:

http://www.mfjenterprises.com/Product.php?productid=MFJ-2912   Isolation balun to insert in coax lines to
    reduce common mode unwanted RF currents
http://www.mfjenterprises.com/Product.php?productid=MFJ-911H   Coax to balanced line balun, switching
    between 1:1 (no impedance change) or 4:1 (4:1 impedance transformer)

NOTES

7 MOVING TRAFFIC & TEACHING SKILLS IN ARES NETS

by Joe Bassett

Effective emergency communication is about one thing: communication. It starts with communication. Communication is at its center. And ultimately it's about communication. This is why the word "communication" is in the term itself. Furthermore, communication of no value if it isn't effective. Ineffective communication is a *non sequitur*, it's noise.

Related to amateur radio emergency communication, radio waves should be utilized to communicate information that results in an effect, i.e. necessary action initiated via a Skywarn report, deployment of needed supplies in the wake of a crisis, comfort in the form of a Safe-and-Well message. Whatever the message, via whatever mode, emergency communication should render an effect.

In that light, amateur radio operators are most effective as communicators who happen to use radios, rather than radio operators who happen to communicate. Communication is the goal, radios are the tools.

The decision to use a radio to send a message is too often answered with the question, "Can I send this by radio?" But effective communicators ask themselves, "*Should* this be sent by radio?" followed by, "Is radio the best means to convey this information?"

Of course, amateur radio has prided itself on the premise of "When all else fails…" Certainly, if all other forms of communication have failed, then use radio waves is warranted. At that point, radio is the appropriate vehicle and amateur radio operators are often the "cavalry" saving the day!

And, yes, there is still the possibility, probability, even eventuality, that all else *will* fail. This is a lesson learned through the loss of 911 capability in local municipalities *and* in large scale disasters such as hurricane Maria in Puerto Rico.

Here in lies the tension between the amateur radio operator's affinity for radio technology, the commitment to serve society, and the pride of proficiency. Effective amateur radio operators balance love for the hobby, drive to proficiency, and a desire to be part of something bigger than themselves.

First, love for the hobby is best revealed in the confession that we hams, if we're honest with ourselves, are still big kids playing with walkie-talkies in the woods. It's imperative to remember this while recruiting and training amateur operators toward emergency communication. Picking up a microphone and speaking information to a listener at the other end is the most intuitive form of radio communication.

Second, many amateur radio communication groups lament a difficulty in recruiting and assimilating new members. At the same time, there is a growing insistence that digital modes of communication be the foundation for supporting disaster response. It is true that digital communication is accurate, however it may not be as effective for the new ham. To illustrate this, we can ask the following question of ourselves… "How many of us began the hobby of amateur radio to talk on the radio versus how many of us began it to type on a computer?"

It's safe to say that the former is true rather than the latter. It's no wonder that the attrition rate of "newbies" is high in many areas. The new ham is most likely attracted by personal interaction and a desire to contribute to a team, many while monitoring emergency amateur communication as scanner enthusiasts. Then we initiate them into service by insisting that they expend additional capital, time and financial, toward impersonal digital modes, before they understand the underlying principles of emergency communication. (Look no further than the lament surrounding the rise of impersonal interaction brought about by texting and social media.)

Third, as stated earlier, digital communication is accurate, but it is less robust. This is evidenced by the fact that it introduces more failure points at both the transmission and reception location. These failure points include both equipment and personnel. When additional equipment is introduced to a signal flow, potential mechanical and electrical failure is increased. Also, additional equipment requires additional training and documentation.

This is not to say that digital modes don't add significant value to emergency communication. To say so negates the truism that amateur radio emergency operators are communicators first and radio operators second. This author has stated publicly that "…amateur radio operators who neglect digital modes as valuable tools for emergency communication do so at the peril of dereliction of their duty during emergency situations." The point is: digital modes are applicable and effective after the basics of radio communication are understood and practiced. These basics are best learned through the practical application of voice communication.

Finally, in the case of digital communication, when a point of failure raises its head, *and be assured that it will*, the skillset of voice communication will acquit itself admirably. With that, amateur radio communication begins and ends with passing messages via voice. By extension, so should training.

Such training hinges on three interdependent components: directed net procedures, message handling (transmission and reception), and message formatting. In turn, each component rests on interdependent skills and disciplines.

Directed Net Procedures

- Net preamble
- Net control procedures
- Check in protocol
- Traffic listing
- Phonetics
- Pro-words and break tags

- Relays
- Communication routing
- Interference

Message handling
- Expediency and accuracy
- Precedence
- Directing traffic
- Copy speed versus reading speed
- Fill requests and provision
- Acknowledgement
- Relay, guarding, and forwarding
- Message retention

Message formatting
- Pro forma
- ICS forms
- Served agency's forms
- Pertinent and actionable information

The application of these disciplines provides and environment as conducive to effective message transmission as does digital communication. In many cases, messages can be transmitted and received/acknowledged with more expediency than digital modes. However, there are myriad instances when digital communication is preferable to voice. These include, but are not limited to: lengthy forms, messages with no time value, and recurring messages. Additionally, many digital modes of radio communication succeed in conditions which render voice communication inoperable.

Once the voice skillset is assimilated by the amateur operator it is easy to apply that skillset to digital modes of radio communication. By extension, the understanding of voice procedures undergirds digital communication. Thus, many options for success are available to the well-rounded and equipped amateur radio operator when all else fails.

On a personal level, the author has found several beneficial maxims for training amateur radio emergency communicators...

Practice doesn't make perfect; perfect practice makes perfect.

If you think that you're going too slow, slow down.

Professionals built the Titanic, amateurs built the ark.

Amateurs practice until they get it right. Professionals practice don't get it wrong.

Jack of all trades, master of one.

When crisis happens, do the best that you can. You can improve later.

8 SOLAR AND OTHER POWER SYSTEMS

Providing alternate power sources in the event of a loss of normal utility power is somewhat just an exercise in General Class power and ohms law equations. Let's review some important numbers to keep in mind:

RADIO EFFICIENCY: Most solid state transceivers use relatively little power in receive, and use a lot more power in transmit. Generally, they will be around 50% efficient in transmit --- to get 50 watts out to the transmission line, they are going to use 100 watts of power from a battery or other source.   Vacuum tube gear is going to be less efficient due to having to run all those filaments. I think I one measured my 100-watt output SB-102 and it used about 170 watts in receive and about 350 in transmit.   An equivalent solid state rig might run 15 watts or so in receie and 200 watts in transmit.

Most of the time you don't transmit all the time, so your "duty cycle" is maybe 30-50%

GENERATORS: Even a vacuum tube transceiver is a very light load for almost any generator.   Camping style generators may put out only 1000 watts where larger ones may be capable of 2, 3, 5, or more kilowatts. Generators require a certain amount of fuel just to rotate, whether they provide power or not, and in general, the efficiency of internal combustion engines is poor – 20%?   So most of the energy of your fuel is going to go into heat.   Figure about 5 kilowatt-hours per gallon of fuel, if the generator is running at its MOST efficient point (and that isn't idling).   Running a 5kw generator at idle to power a radio that averages about 60 watts is going to be terribly fuel-inefficient.

Diesel has more energy per gallon than does gasoline and the diesel combustion cycle is more efficient than the gasoline cycle. (And small diesel generators are extraordinarily rare and very expensive.)   Propane has less energy per gallon than does gasoline.

Fuel	BTU/gallon
Propane	91600
Gasoline	114000
Kerosene	134000

BATTERIES.   A recently purchased marine deep cycle battery might have a capacity of 75-100 Ahr. At 12VDC, that is 900 – 1200 Watt -hours.   You probably don't want to discharge it more than 80% of its capacity, but that still gives you about 700 watt-hours – or 12 hours of usage if your average usage is 60 watts for a heavily used 100-watt output solid state rig.

A little 7.5 Ahr 12v gell-cell would have about 75 watt-hours of usable power in it, but if you're running a 20-watt output vhf transceiver (40 watts dc power when transmitting) and your transmit time is minimal, you might have an average power requirement more like 15 watts and get away with 5 hours of usage off a relatively small $12 battery from https://www.apexbattery.com/

Battery Capacity	
7.5 Ahr 12V $12 Sealed Lead Acid	https://www.apexbattery.com/12v-7-5ah-f2-sealed-replacement-battery.html
10 Ahr 12V $19 Sealed Lead Acid	

NOTE ON LEAD ACID BATTERIES: There are precise recommendations for charging and maintaining lead-acid batteries depending on type (flooded lead acid; maintenance free; gell cell etc all take slightly different voltages).   In general, battery life is a tradeoff between sulfation due to too low a stored trickle charge voltage and corrosion due to too-high.   High quality batteries (such as Rolls batteries for solar power installations) come with exact recommendations and charge controllers go through multi-phased careful charging cycles.   For home use of typical lead acid batteries, the most important considerations are: recharge the battery on a regular basis (e.g. monthly) to deal with self-discharge and avoid sulfation, and avoid over-discharging a battery, particularly one not meant for "deep cycle" use.

USING YOUR CAR: You can take advantage of the alternator in your automobile or truck to provide 13.8VDC; alternators can produce as much as 75amps ifyou connect right to the battery (10A typical limit due to fusing of the cigarette lighter) --- but beware, that big 100 HP engine is going to waste a good bit of fuel!

SOLAR.   Solar panels come in many different sizes and voltages.   The smaller panels intended to charge 12v batteries typically produce about 18V in direct strong sunlight and require some sort of "charge controller" or they will cook/fry/destroy a battery by over-charging it eventually.   Larger panels intended for home or commercial power generation often come in 30VDC open circuit voltage.

Solar panels generate less voltage when they are hot, and more voltage when they are cold.   That's why home installations typically produce more POWER in spring / fall than in summer, despite having more hours of sun in the summer.

Charge Controllers: There are two main technologies for charge controllers:
- Pulse Width Modulation (PWM) rapidly switches solar power on and off to the battery (or load) so that the average voltage is correct. (If the panel is much higher voltage than the battery, the extra voltage has to be lost in the panel/wiring resistance, decreasing efficiency).
- Maximum Power Point Transfer (MPPT) controllers are significantly more expensive, but perform a true dc-dc conversion (usually with an oscillator in there, and a switching power supply) and choose the load they place on the solar panel so as to operate at the voltage at which the solar panel is MOST efficient --- they may be as much as 30% more efficient than a PWM controller (according to their advertisements) and this make a difference for continuously used home installations.   Casual portable emergency systems probably will utilize the cheaper PWM controllers, but this requires that you have panels that are not far above the voltage of your batteries for reasonable efficiency.

Figure. Inexpensive MPPT controller (max 15 amps output, max 60V input, auto detect and configure for 12 or 24 volt battery system) that can allow usage of 30V nominal panels with 12V battery system and good efficiency.

Probably more important than the type charge controller on your system is AIMING THE PANELS: You can dramatically increase the power produced throughout the day by your emergency solar panels if you will re-aim them every hour or so to keep them aimed exactly at the sun in 2 dimensions.

Costs: While larger panels are down below $1/watt (DC), you'll be lucky to get smaller systems at $2/watt. My friend Art Grant recommends a 100-watt package from Amazon that includes a pwm controller and has everything except the battery in a nice carrying case also:

https://www.amazon.com/ACOPOWER-Foldable-Generator-Suitcase-Controller/dp/B01MU4PGLQ/ref=sr_1_2?ie=UTF8&qid=1516597980&sr=8-2&keywords=acopower+solar

Solar Battery Backup: Because you might not have all the sun you wish, you probably wish to have a somewhat larger battery, so that you can easily make it through the "dark hours" and then through a cloudy day the next day (or more!) for emergency operations. So consider what the power level of your transceiver needs to be and purchase an appropriate amount of battery backup! In a pinch, you might use regular automotive batteries, but remember they are typically made for quick power, not deep discharge--- try to avoid discharging them more than 50% of their rated capacity, and remember that used car batteries may no longer have their "rated" capacity.

Be sure to use BIG WIRES if you are running things off 12 volts. Otherwise you're going to lose a lot of energy merely in the wires --- and possibly even have an overheated wire. Remember that all power sources-- particularly batteries – need a fuse RIGHT NEAR THE SOURCE --- the goal of the fuse is to burn out BEFORE the wiring catches fire!!! Never fuse a circuit at a higher amperage than is safe for the wiring you're using.

Typical Amperage Capacity of Wiring:

Wire Size	60 deg C rated wire (NM-B, UF-B)	75 deg C rated wire (THWN)
14	15	15
12	20	20
10	30	30
8	40	50
6	55	65

REFERENCES:
https://settysoutham.wordpress.com/2010/05/26/portable-generators-about-half-as-efficient-as-power-plants/

NOTES

_____
_____
_____
_____
_____
_____
_____
_____
_____
_____
_____
_____
_____
_____
_____
_____
_____
_____

9 DIGITAL TECHNIQUES FOR HEAVY HAULING

One way of looking at the types of communications required in an emergency is to break them up in a 2x2 box by looking what the nature of the information to be transferred (tactical versus logistical) and the nature of the connections required (broadcast versus 1:1):

Nature of the information: Nature of the connections:	Tactical Communications – short pithy exchanges of information that don't have to be recorded word for word.	Logistical communications: detailed "record" traffic that may include long sets of data that must be delivered and recorded precisely without error.
BROADCASTS – need to be sent to many locations at the same time	VOICE – VHF FM (short distance) or SSB (long distance) In difficult propagation circumstances, PSK31, JT65 or other techniques might be utilized.	PK31 / MT63 / Olivia and other digital techniques come the closest.
1:1 – need to be sent to just one recipient		WINLINK – either via gateway or peer to peer (error free) FLDIGI/FLMSG – error free YAPP via packet – error free

Amateurs have typically excelled at the TACTICAL/BROADCAST solutions, primarily VHF Voice within a city, and SSB Voice to pass information long distances. Both played very important roles in Puerto Rico. The "Force of 50" was sent ostensibly to do one of the lesser-mastered roles in amateur radio: 1:1 / Logistical communications to fill a Safe&Well database (for which WINLINK was selected).

Everyday Communications
Cell phones handle civilian Tactical communications quite well, and with voice conferencing they even allow for "broadcast" techniques.   Web-conferencing also fills that need. Text messaging however has taken a huge bite out of 1:1 tactical communications!

Email is used by the general public to handle the detailed communications, and to some extent text messaging can fill in there also with photos etc.

Well Rounded Amateur Radio
The expert amateur radio emergency communicator needs to have skills in all quadrants.   Some groups espouse "lowest common denominator" communications, but this ends up with amateur radio unable to meet the need until the military arrives with satellite dishes and begins to handle the detailed logistical comms required to handle tens of thousands of deliveries, and personnel. If your ARES or other emergency group misses out on the record logistical comms techniques, you're going to be limited in a larger emergency....

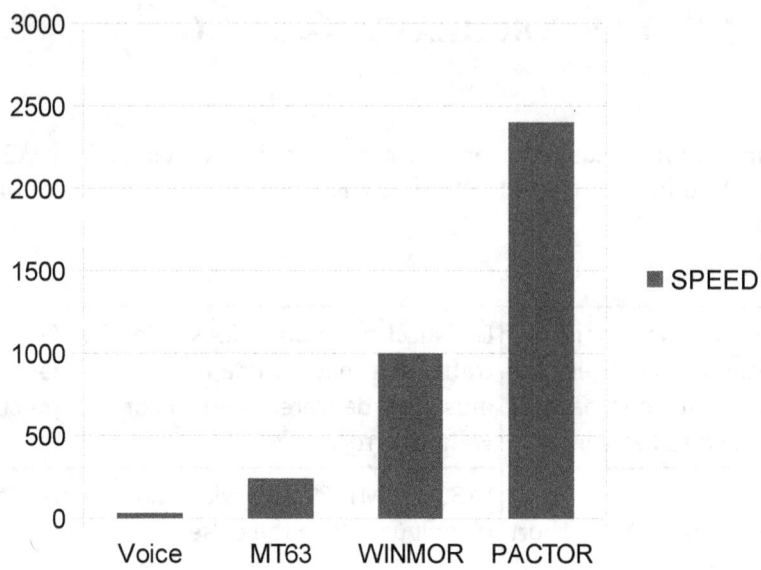

Figure: Digital techniques vastly outpace voice when larger files or error-free transmission is required. Speeds are for strong signal environments, characters per minute.

DIGITAL TECHNIQUES

Familiarity with FLDIGI/FLMSG, and WINLINK will fill most of the need for detailed error-free (ARQ, handshake) communications; adding packet YAPP gives yet another technique.   These techniques have vastly higher throughput for long messages than does voice, primarily because record voice must be TRANSCRIBED and that is generally limited to 20-30 wpm, whereas digital techniques can easily reach 1000-2000 characters per minute (or, assuming 5 characters per "word", 200-400 words per minute).   Directly transcribed onto screen or computer file, they can run rings around voice transmissions in good propagation, and newer techqniques can even deal with difficult propagation conditions.

Your group would be wise to master these techniques.   You can start with any of them, and work your way through them.   Software for all three is free and not difficult to learn as evidenced by hundreds of thousands of downloads of FLDIGI software, and a hot market for sound-card interfaces that allows prices of simple circuits to rise above $100.

Note: Sarasota Digital Group has excellent materials: http://n4ser.org/sdg/#Tutorials

WINLINK

Download site: on www.winlink.org

ftp://autoupdate.winlink.org/User%20Programs/Winlink_Express_install_1-5-10-0.zip
(> 20 Mbyte installation file)

For HF you'll also need a propagation program written for the US government to do HF broadcast propagation predictions know as "itshfbc.exe":   Greg Hand has the best download site:
http://www.greg-hand.com/versions/itshfbc_161207.exe

For VHF packet, you'll also need either a hardware TNC (e.g Kantronics KPC-3) or a soundcard/soundcard interface and some modem program – most of us use UZ7HO's free soundmodem.exe:
http://uz7.ho.ua/modem_beta/soundmodem97.zip

While you are there, download his excellent plain-jane terminal program easyterm:
http://uz7.ho.ua/apps/easyterm39.zip

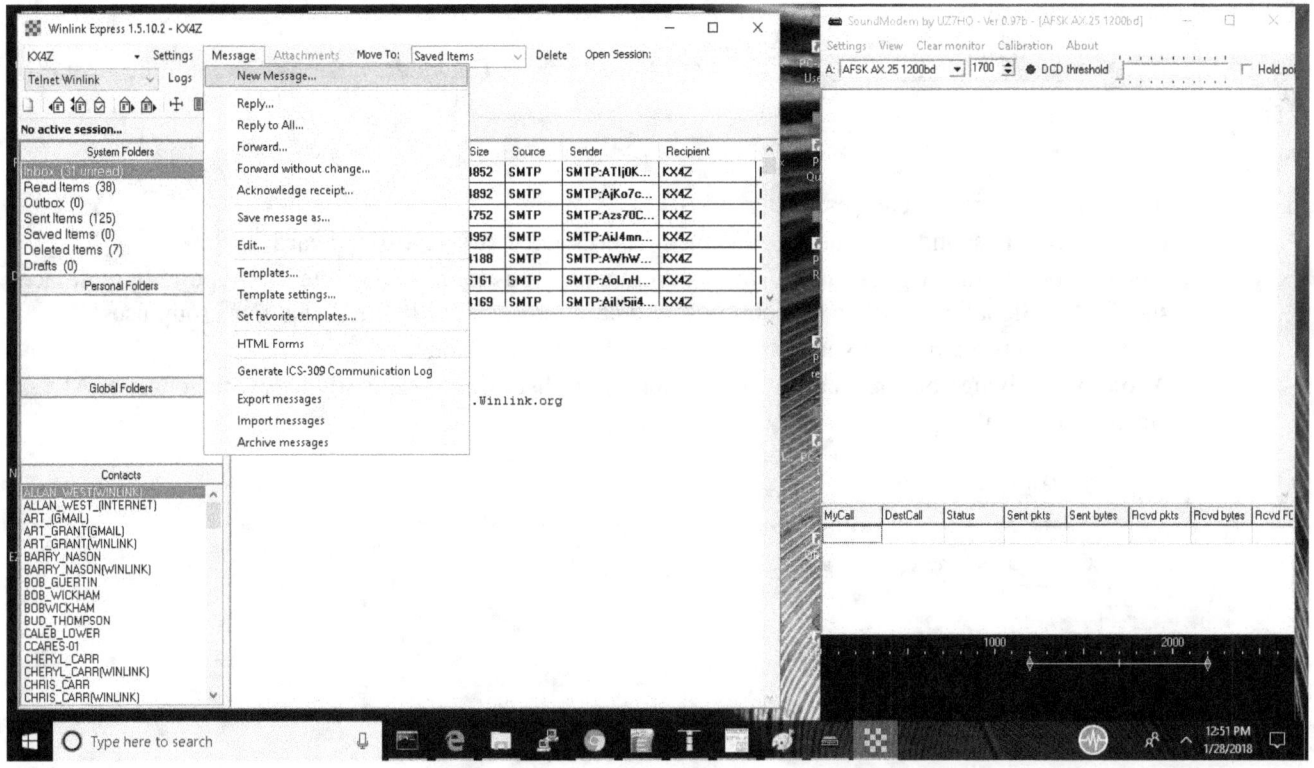

For HF sound-card usage (WINMOR-WINLINK) all that is required is to download an updated list of gateways ("channel selection") --and set the TX and RX gain on the signalink so you can see signals and aren't overdriving your transmitter.   Use the test transmission tone to set gains.

For VHF soundcard usage (packet, using soundmodem97.exe) you'll have to work your way through some of the configuration of soundmodem97....this can be tricky.   Best to have a mentor, but we have instructions here:

http://www.qsl.net/nf4rc/UnderstandingAudioChannelConfiguration.pdf

FLDIGI

Base program (with the modem protocols) is fldigi.   Flarq uses these to send ARQ files and/or text; this corresponds to sending email attachments in WINLINK. . Note that Olivia cannot be used. Flmsg corresponds to templates in WINLINK – has the ability to error free send all kinds of message formats.

Primer on just fldigi: http://qsl.net/kx4z/FLDIGICheetsheetforVolunteers.pdf

Download site:
    FLDIGI   https://sourceforge.net/projects/fldigi/
    FLMSG   https://sourceforge.net/projects/fldigi/files/flmsg/   (includes versions for multiple operating systems)
    FLARQ:

FLDIGI – very busy configuration choices.
- Pay attention to sound card choices.   Typical "port audio" & select sound card.
- Pay attention to the "right channel" for sending continuous tones to activate vox-type PTT in devices such as the Signalink (and the homebrew sound card interface "$25TNC" that our group uses
- Learn how to recognize the various sounds of different digital signals
- Avoid overdriving your transceiver into compression/distortion – generally below the onset of ALC
- No compression!

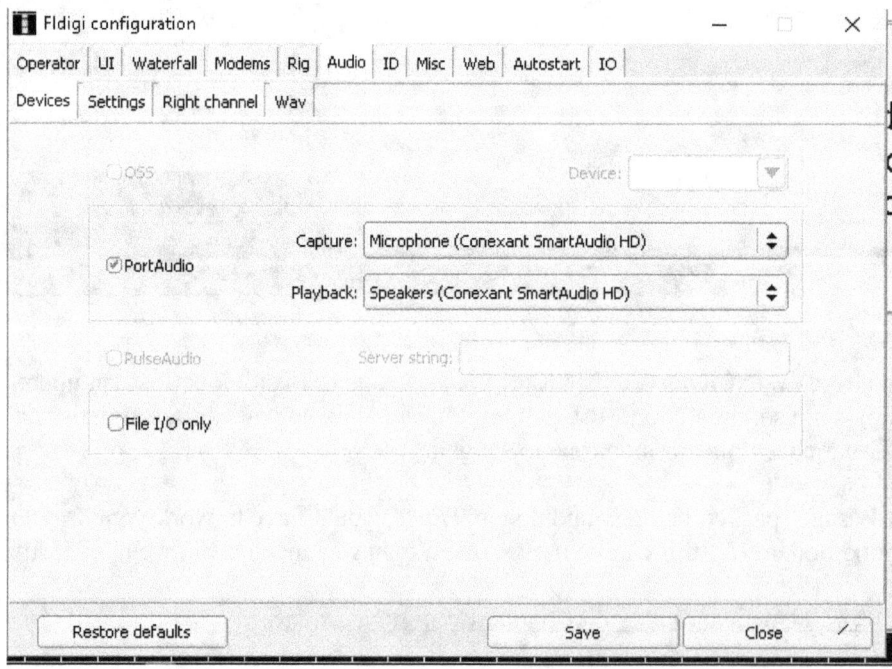

Figure – here the internal soundcard of the laptop is being selected to demonstrate sounds – but typically you'll be selecting some "USB CODEC" or similar.

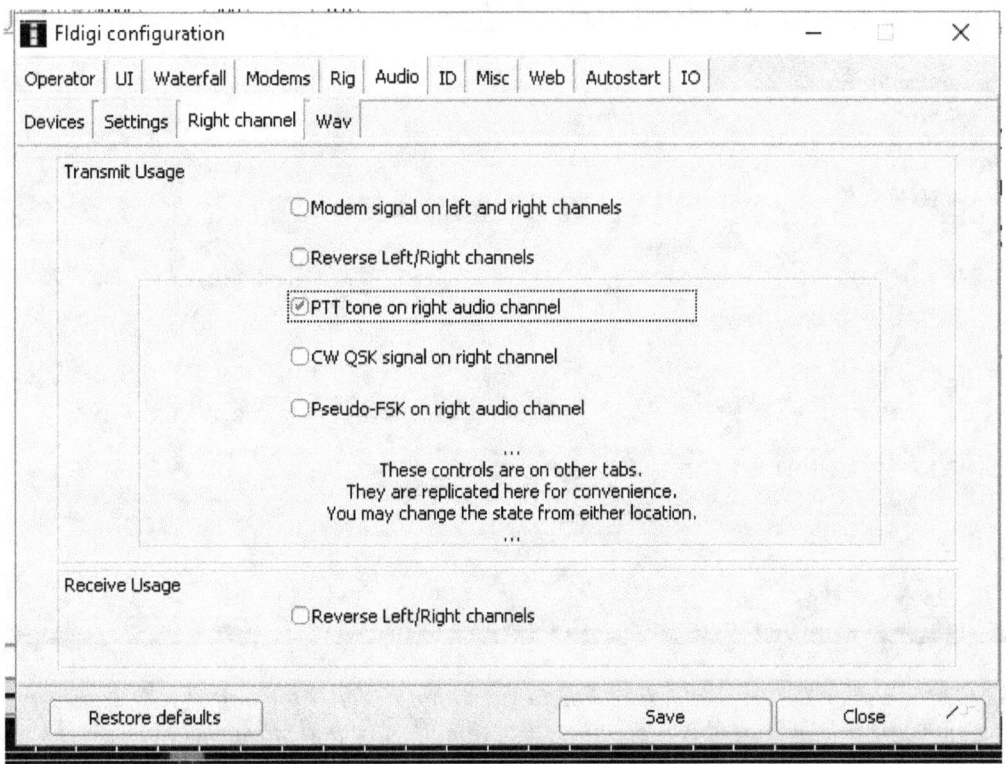

Figure – typically you will want the PTT on the right channel (versus left) to actuate signalink or similar vox-type PTT soundcard interfaces --- but yours just might be wired backwards!

OPTIONAL RIG CONTROL:

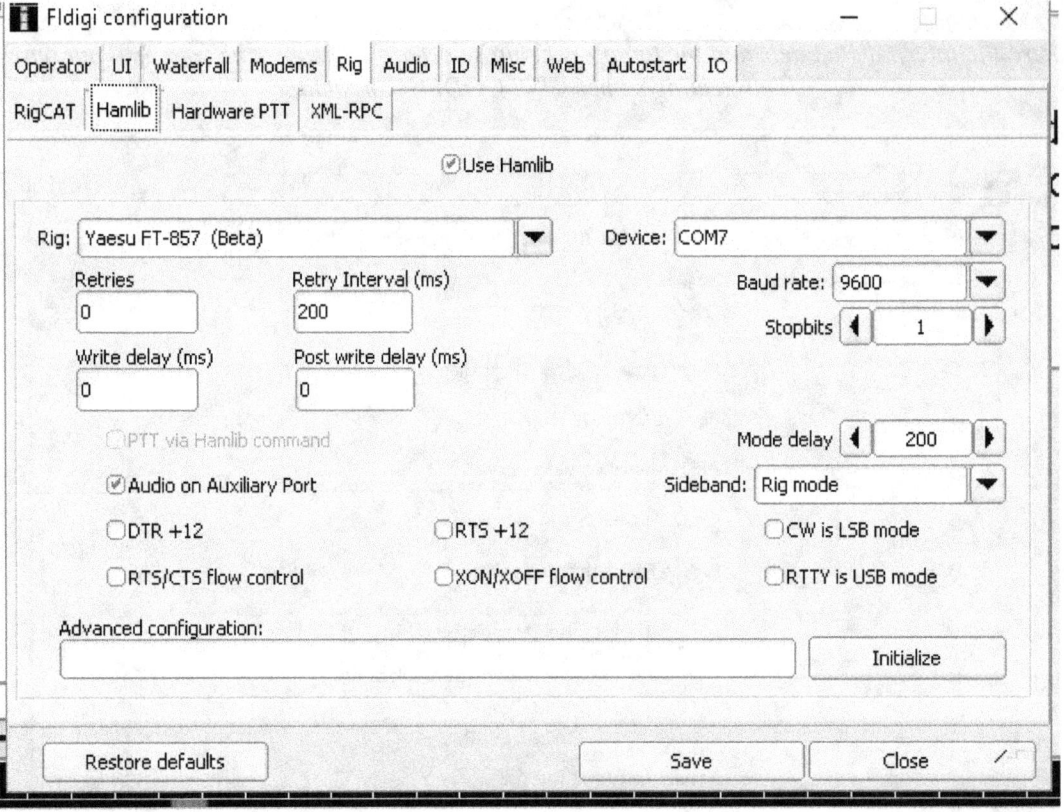

Figure – RigCAT, Hamlib and other methods are included within FLDIGI to control your transceiver's frequency (and more) if you desire – this is completely optional!   I've had more success with HamLib.

FLMSG:

When loading flmsg – be certain to select the "expert" interface; the "served agency" interface doesn't do what you need.

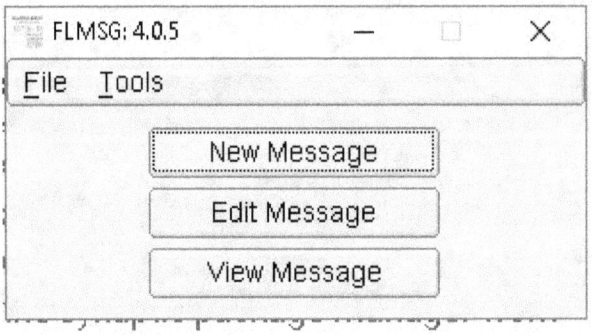

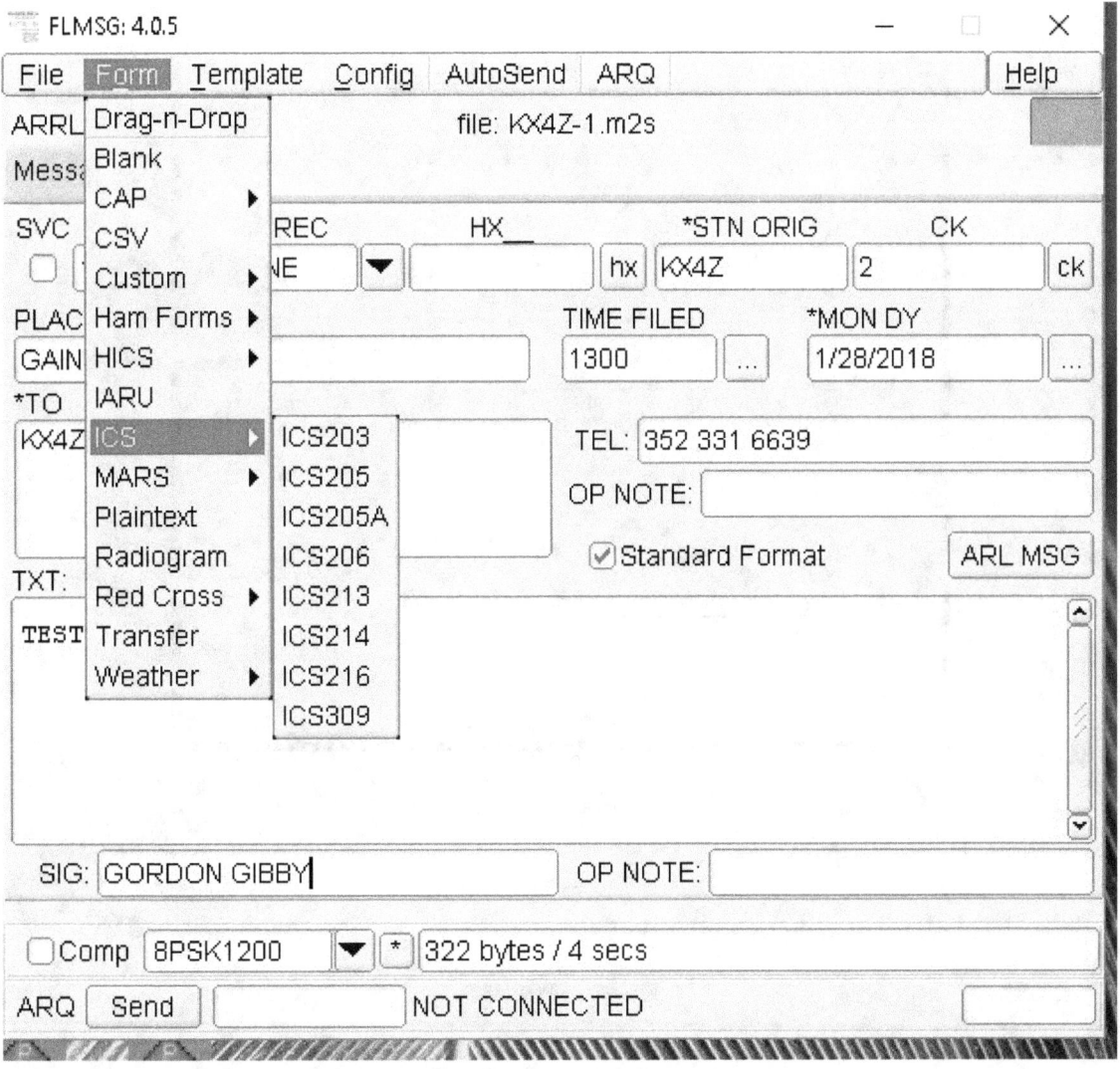

EASYTERM/ YAPP

Not to be outdone, the free terminal program EASYTERM has a "YAPP" protocol that can do unattended error free file reception – quite useful.

http://www.qsl.net/nf4rc/EasyTermTutorial.pdf

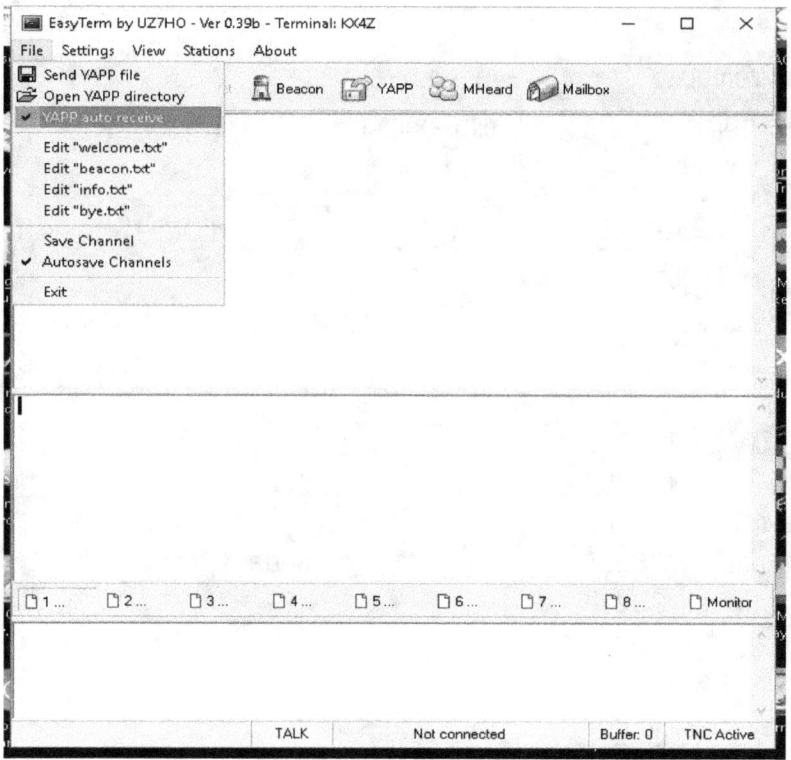

10 WORKING WITH YOUR EOC

by Jeff Capehart W4UFL

ASSUMPTIONS

We assume you have already established a relationship with your local county Emergency Operations Center (EOC) and Emergency Manager. The State of Florida Division of Emergency Management (FDEM) has selected the ARES program as the preferred volunteer communications group to coordinate with for Emergency Support Function (ESF) 2 – Communications. If your local group has not established a relationship with your local EOC, then by all means, begin work to make your group known, active, and of service.

ROLE OF THE PROFESSIONAL VOLUNTEER

To build and maintain a good, working relationship with the EOC there are many factors to consider doing as well as what not to do.

- Get involved BEFORE an emergency happens. Take classes. Participate in training.
- Learn what the role of the EOC is and how you will fit in. Be polite.
- Take the required ICS courses and keep copies of your certificates.
- Dress properly. Follow the rules. Be consistent and reliable. Show up when asked.
- Get the background check and badging done.

AMATEUR RADIO EQUIPMENT

You will need working amateur radio equipment at the EOC in order to do the best job of emergency communications support. If your radios, feedline, or antennas are not up to the job, or non-existent, then it is time to install or upgrade the equipment. This can be especially problematic when dealing with government bureaucracy such as multiple-agencies, budgets, purchasing departments, facilities staff, and professional tower climber crews. Some factors to consider:

- Is FCC "Type Certified" commercial radio equipment required?
- Do the EM staff understand the value of HF, VHF and UHF radios?
- Where do you begin to get faulty antennas and ill-planned systems get fixed?
- What are the right political ways to work with the system?

- Competition with public safety for space on the radio tower?
- Does the building have backup power systems?
- Are outside computers (laptops, ipads) allowed on the network or guest WiFi?
- Is a computer provided and how can hams get specialized software (like Winlink FLDIGI) installed and maintained?

WORKING WITH THE EOC STAFF

Key steps to help smooth a working relationship with your EM staff include the following:

- Understand the politics of the department
- Try to work with the personalities or figure out how best to work with them
- Be understanding of their very limited manpower
- Find ways to make problems more clear what things need fixing
- ✓ Conduct a performance study of equipment and write a report
- ✓ Conduct exercises that stress the ability of the equipment to perform
- ✓ Benchmark with other EOC's to see what their peers are doing

Take things up a notch when you aren't getting anywhere.   For example, if your EOC is run by the Sheriff, then find out what they would do if they lost their radio system. Explain how the amateur radio system can help, but only if the right equipment is available and working. Find the key people and have meetings with them.

THE ARRL EMERGENCY COORDINATOR MANUAL

There are several chapters in the EC manual that provide insight and direction for new Emergency Coordinators. Take some time to read the manual and take advantage of hams who have done this before so that you don't have to re-invent the wheel and learn everything on your own the hard way through trial and error.

1. How the Government Sees You
2. ARES v. RACES
3. Developing an Emergency Communications Plan
4. Being Active - Participating in Training
   Get ICS Training
   Meet the "EOC Qualifications"
   Make yourself a "known quantity"
   Demonstrate that you are reliable
   Be familiar with their tools like WebEOC
   Show up for exercises and table-tops
   Provide a service of value (i.e. check school weather radios)

5. Reporting to ESF-2 under ICS
6.   Coordinating Volunteers

Selling the Agencies on ARES

<u>Table 13-1 General Format of Your Presentation</u>
1. Formal Introduction.
2. Brief explanation of your duties and responsibilities.
3. Brief explanation of the ARRL Field Organization.
4. Statement of Purpose.
5. Demonstration (handheld, video, etc.).
6. Question and answers.
7. Comments about your local ARES group.
8. Determine the agency's needs.
9. Leave information.
10. Schedule second appointment.
11. Thank the administrator.
12. Leave.

<u>OPEN DISCUSSION AND Q&A</u>

NOTES

11 EMAIL TECHNIQUES - VHF

by Vann Chesney (AC4QS)

Why Email?
Everyone has a VHF handheld or mobile – who needs email?

Software
RMS Express - www.winlink.org
For VHF packet - uz7.ho.ua/packetradio.htm

Hardware - PC
Windows computer.
At a minimum, the computer should be a modern 500 MHz or greater Pentium/Celeron running one of the following OS with .NET 3.5 and/or 4.0 installed:
- Windows Vista (32 bit or 64 bit versions)
- Windows 7
- Windows 8
- Windows Server 2003 or later

Hardware - TNC/Sound-card Interface
TNCs - Kantronics, TNC-X, MFJ etc.
Sound-card Interface - Signalink, Rigblaster, Homemade, etc.

Live Demo
Description of equipment and connections between computer and radio.
Winlink Software
- Setup
- Configuration
- Main Window
- Session Types
- Radio Interface Setup
Soundmodem Setup
Send Demo Email

NOTES

12 EMAIL TECHNIQUES: HF

by Gordon Gibby KX4Z

WINLINK is the defacto standard for amateur radio email at the current time, with thousands of users, processing scores of thousands of emails every month. On the HF bands, there are 50 or more volunteer gateways in North America and a smattering more throughout the world, providing communications free of charge to mariners on the high seas, people in desolate places, and anyone else.   By securing a base set of users (boaters on the high seas) WINLINK has garnered a significant "edge" over other traffic management systems which typically have to work hard to find material on which to have continuing practice.

WINLINK's weakness is that their client software is somewhat limited to Windows operating system (though some intrepid souls have gotten it working on Linux or Apple products with effort) – and both Windows itself, and the WINLINK software continue to "update".....which can be a problem when a new incompatibility or bug is introduced.

Nevertheless, anyone involved in emergency communications probably needs to gain intimate familiarity with at least the client (user) side of WINLINK, and probably would be a stronger participant with some knowledge of the server (gateway) side as well.

Software can be downloaded here:
 ftp://autoupdate.winlink.org/User%20Programs/Winlink_Express_install_1-5-10-0.zip

You will also need a free government-developed propagation software (originally for predicting how best to receive broadcast stations) *itshfbc.exe*   which can be downloaded here-- please let it install in its preferred directory –   http://www.greg-hand.com/versions/itshfbc_161207.exe   This software is utilized in a nifty scheme to direct you to server stations on bands that you're most likely to succeed.

REGISTRATION
WINLINK software may ask you for a "registration" number – merely wait a few moments and it will allow you to ask for a "remind me later" and allow you to proceed.

Email programs always have some method to create messages, send messages (usually just push SEND); read messages, delete, forward, reply, etc.   WINLINK has all of these features as well as a rudimentary contact list, but there is a crucial difference:

Normal email programs have only ONE method to transmit—they use the internet-- so they merely provide a SEND button.   WINLINK has MANY methods you may transmit messages so there is an intermediate step of "posting to Outbox" and then you must choose and open a "Session Type" and establish a connection to transact emails.

Some of the Session Types and their explanation follow – these are important to understand!

SESSION TYPE	USAGE	COMMENT
Telnet Winlink	Uses the Internet to make a connection to a Winlink Central Message Server (CMS) to transact email similarly to normal Internet email programs.	
Packet Winlink	Utilizes AX.25 Packet communications (typically used on VHF/UHF) to connect to a Packet gateway.	WINLINK does not include the modem protocol – you must utilize a hardware TNC (e.g., Kantronics) or alternatively a sound-card interface and a program such as soundmodem.exe from UZ7HO (quite popular)
Pactor Winlink	Uses an external hardware PACTOR modem (marketed by a private company, SCS) to make high speed error free connections on HF bands.	Throughput reaches about 2000 characters per minute; a favorite of boaters on the high seas. Costs are > $1000 but that's the same range as some people pay for their transceivers these days. Works really well!
Winmor Winlink	Built into WINLINK – utilizes a soundcard interface to achieve medium speed connections, typically on HF	The "nagware" will continue to ask you for a donation but allows you to proceed. It is a worthy group and your donation is worthwhile.
Packet P2P	Allows you to transact email directly to your recipient over packet, without needing a gateway station.	Useful if you have confidential information that you do not want accessible through the normal WINLINK administration.
Winmor P2P	Allows you to transact email directly to your recipient over WINMOR, without needing a gateway station.	Useful if you have confidential information that you do not want accessible through the normal WINLINK administration.
Winmor Radio Only	Initiating an email in this system forces it to follow the pathways utilized when there is no Internet – a much slower backup system built into the WINLINK radio network. Delivery is not guaranteed, and the recipient must have selected a "Message Pickup Station" for this to succeed.	Useful way to test how messages would work if the Internet were not available at all.

On HF, you must select a radio station to try and contact.   Figuring out what band to use and where are potential stations can be confusing at first – the WINLINK software includes a "Channel Selection" item in the menu bar of radio-type sessions which allows you to easily find good choices.

HF Channel Selector   ✕

Exit   Select   Update Table Via Internet   Update Table Via Radio   Forecast   SFI   All RMS   ▼

Callsign	Frequency (kHz)	Mode	Grid Square	Hours	Group	Distance (km)	Bearing (Degrees)	Path Reliability Estimate	Path Quality Estimate
KX4Z	10141.000	1600	EL89RQ	00-23	PUBLIC	0	000	100	100
KX4Z	3595.500	1600	EL89RQ	00-23	PUBLIC	0	000	100	100
KX4Z	18106.200	1600	EL89RQ	15-23	PUBLIC	0	000	100	100
KX4Z	7104.000	1600	EL89RQ	00-23	PUBLIC	0	000	100	100
KX4Z	14098.700	1600	EL89RQ	00-23	PUBLIC	0	000	100	100
WD4SEN	3587.500	1600	EM90CC	00-23	PUBLIC	85	058	88	52
WX4PCA-10	3591.000	1600	EM73NU	00-23	PUBLIC	512	335	88	50
KJ4EWA	3586.500	1600	EM80IW	00-23	PUBLIC	156	333	86	51
WW4MSK	3592.500	1600	EM74UW	00-23	PUBLIC	606	345	86	49
N4JGW	3597.000	1600	EM74LR	00-23	PUBLIC	607	338	86	49
KI4WPI	3598.500	1600	EL96UF	00-23	PUBLIC	443	150	82	49
AK4SK	3570.000	500	EM60VL	01-12	PUBLIC	364	285	81	49
AK4SK	3591.000	1600	EM60VL	01-12	PUBLIC	364	285	81	49
KW4PD	3593.500	500	EM95UF	00-23	PUBLIC	651	018	80	47
KW4PD	3593.500	1600	EM95UF	00-23	PUBLIC	651	018	80	47
K8KHW	7104.500	1600	EM99JJ	00-23	PUBLIC	1085	006	78	47
KQ4ET	7101.200	1600	FM16XU	00-23	PUBLIC	998	035	77	47

Just like a spreadsheet, this popup can be sorted by any of its columns.   I prefer to sort of "Path Reliability Estimate" – click this until the best one appear on the top.   Then it shows you the stations you're most likely to be able to connect with, their frequency, mode, distance, and bearing.   Double click on a desired station and it will return you to the connection dialog.   If you have your transceiver connected for computer frequency control, it will actually tune you for the correct frequency!

A word about propagation:

Nightime winlink connctions are generally easy.   On 80 or 40 meters, you are likely to have a wide range of possible choices.   That continues until about 2 hours after sunrise.   But during the hottest portion of the day, the solar insonation of the D layer is most intense, and lower-frequency bands (like 80 meters) become fairly unusuable....your choices just after local noontime may be quite limited for possible connections.

If you're planning a full scale exercise you may want to get on the air a few days beforehand and test out some possible gateway stations as not all stations are identical.   Some may have wonderful antenna arrays and higher power transceivers – others may be quite modest stations and have a weaker signal, or be in the midst of broadcast station interference.

HOW TO GET A WINLINK ACCOUNT / EMAIL ADDRESS

You must use the downloaded software to create an account.
It is much easier to do this BEFORE the incident, while the Internet is still working.
Go to the Setup Page, enter all the information, make up a Password, and hit UPDATE.
That creates an account for you.   Your email address is   <yourcallsign>@WINLINK.ORG

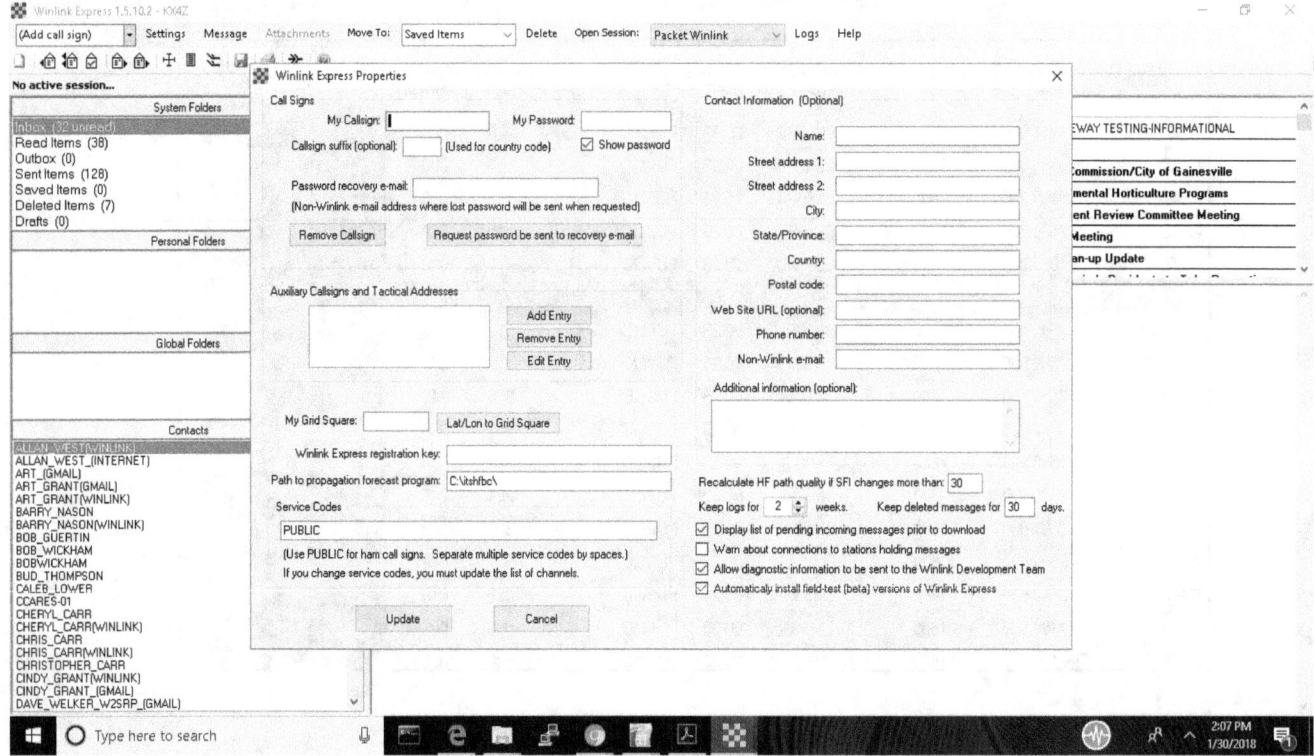

The following procedure is suggested if you are forced to create an account with only radio access:

Connect with the system (send a message) to create your account. Do not use a password on your first connection. Your radio email address is YOURCALL@winlink.org. A message containing your password will be sent to your account. Retrieve it with a second connection. Secure login will now be enforced by the CMS so, be sure to set your password in your client program.

1.At any time, use the form (lower right on the 'My Account' page) to have a password re-sent to your account address. Retrieve it using your client program.
2.Go to 'My Account' from the menu. Use your callsign as your username, and password to log in. You can change your password once you're logged in.

Solutions Worth Knowing...

No.	Area	Details
1	Loss of Internet in local area.	Clients with HF abilities can simply connect to HF gateways in unaffected areas. Servers (gateways) in affected areas will switch to "radio mode" and begin to send email by radio to distant servers with CMS access.....IF THEY ARE PROPERLY CONFIGURED TO ALLOW THAT. Servers ostensibly set to do that are marked as "HF Forwarding" on the winlink map.
2	Loss of internet EVERYWHERE	If there is no internet, you can't use the internet to retrieve email, so you have to retrieve it by radio...But that only works if you were foresighted enough to pick a couple of MESSAGE PICKUP STATIONS (MPS) in your user settings. Messages for you will be mirrored at all the MPS stations you select; pick just 2 as such an occurrence will certainly tax the system!
3	Adding email addresses to your whitelist	The easiest is just to send them a winlink email. That automatically adds them to your whitelist....you might want to do this for anyone you're hoping to have connectivity to in an emergency....or else edit your whitelist online to include them.
4	"Radio-Only"	**Pactor/Winmor Radio-only connections** -- These make radio connections to an RMS that's operating as a node in the Winlink Hybrid Network. Any messages transmitted with this connection are relayed via radio to the destination Message Pickup Stations (MPS) using radio-forwarding from RMS-to-RMS. **Messages do not get sent to a CMS**. This also is the mode you use to pick up radio-only messages from one of your MPS stations. So in other words, this only works if your recipient has a MPS selected....
5	Testing internet outages	RMS RELAY has a setting for "simulate Internet outage" that makes it pretty easy for a local gateway to simulate loss of the Internet. The computer-controlled radio-forwarding works, but it isn't nearly as efficient

		as a trained and experienced human picking stations to whom to connect (been there, done that).
6	Radios feeding into a RMS RELAY	RMS RELAY can accept tcp/ip connections from a variety of radio systems, including RMS_TRIMODE running on the same or other computer (HF transceivers); RMS PACKET running on the same or other computer (VHF AX.25 Packet) ; BPQ32 and linbpq systems also. This allows a multitude of bands and modes to be covered by one gateway system. However, RF Forwarding will ONLY go through the FIRST listed TRIMODE client.
7	PACTOR modems	RMS_RELAY cannot forward through any other type of system other than an SCS PACTOR modem. So it cannot forward through a soundcard (unless that can emulate an SCS PACTOR modem). That's why HF Forwarding gateways have to bite the bullet and get a great modem.
8	WINMOR	Winmor is a wonderful entry level sound-card-based modem protocol that is included in the WINLINK EXPRESS client software. Just about any soundcard-based system (with PTT control) can be utilized. You can also use CAT technology to control many transceivers.
9	Most common glitch with these systems	Upgrades to both WINLINK and Microsoft windows occur frequently....I've had systems fail that got "upgraded" in between being tested and the day of the Exercise....so always keep a spare copy of working INSTALLATION SOFTWARE on a usb drive.

13 MANAGING HOSPITALS
(DEVELOPING A SUCCESSFUL EMERGENCY COMMUNICATIONS PLAN USING HAM RADIO)

Dave Welker, W2SRP
MUNROE REGIONAL/OCALA REGIONAL MEDICAL CENTER
OCALA, FLORIDA HOSPITAL EMERGENCY COMMUNICATIONS HAM VOLUNTEER
COORDINATOR

The Power Point presentation addresses incorporating amateur radio in a hospital emergency communications plan. How the Marion County, Florida hospitals identified the need to incorporate HAM radio. Identifying the goal determined the requirements of equipment and radio operator skills. Examples of developing a program to manage the radio operators include vetting ham operators for volunteer status, (actual ID Badged volunteers), training meetings, exercises, adhering to hospital volunteer policies and requirements such as maintaining a minimum of volunteer hours, completing annual health and safety courses, and developing an understanding of the healthcare facility "culture".

An example of using a MS ACESS database to keep track of volunteers' skills and their availability for "Call up" will be shown.

HAM Radio exercises are based on the NIMS/FEMA guidelines- Hospitals use the Hospital Incident Command System.   Practice with specific HICS forms on a monthly basis using HAM radio equipment ensures their competency to assist in hospitals mandatory exercise requirements for accreditation and for any real disaster. The presentation includes images of hospital exercises, radio installations and an overview of the modes used. There are some examples of after action reports based on the HSEEP (Homeland Security Exercise and Evaluation Program).

References: Managing Hospitals, HAM Radio Emergency Communications- Dave Welker, W2SRP

HAM Look up information to find licensed ham operators in your area, enter the ZIP code in the look up field.
http://www.radioqth.net/ZipLookup.aspx
- **FCC Call sign look up**   http://wireless2.fcc.gov/UlsApp/UlsSearch/searchLicense.jsp
- **HAM Radio Operators lookup by zip code**   http://www.radioqth.net/ziplookup
- **TJC,** The Joint Commission Emergency Management Standard
- **ASPR**- US Health and Human Services, Assistant Secretary for Preparedness and Response

https://en.wikipedia.org/wiki/Office_of_the_Assistant_Secretary_for_Preparedness_and_Response
*   **Order ICS Forms:** NWCG National Fire System Catalog part 2 publication 2013.
www.NWCG.gov/pms/pubs/catalog.htm
*   **HSEEP- Homeland Security Exercise and Evaluation Planning**
   https://www.fema.gov/media-library/assets/documents/32326
●   **FEMA PREPAREDNESS TOOLKIT-** https://preptoolkit.fema.gov/web/hseep-resources
●   **IS 120a Introduction to Exercises**
   https://training.fema.gov/is/courseoverview.aspx?code=is-120.a
*   **Department of Homeland Security case Study on Amateur Radio**
   http://www.dhs.gov/sites/default/files/publications/nppd/Case%20Study_Tennessee%20Violent
%20Storms.pdf

*   **HIPPA and Amateur Radio**   http://hdscs.org/hipaa.html
*   **NIMS On-line training:**   http://training.fema.gov/IS/NIMS.aspx
*   **Florida ICS course schedule:**   http://trac.floridadisaster.org/trac/trainingcalendar.aspx

*   **Difference between Goals and Objectives:**
   http://www.differencebetween.net/business/difference-between-goals-and-objectives/

*   **FEMA radio communications worksheet**   (Assess your communications capability)
   http://emilms.fema.gov/IS775/assets/AssessingRadioCommunicationsCapabilityWorksheet.pdf

*   **The Florida 2011-2013 Public Health and Health Care Preparedness (PHHP) Strategic Plan**
    2.2 Interoperable Voice and Data Appendix A
       Communications - A continuous flow of critical information is maintained as needed among multi jurisdictional and multidisciplinary emergency responders, command posts, agencies and the governmental officials for the duration of the emergency response operation

*   **Centers for Medicare and Medicaid Emergency Preparedness Rule**
   https://www.cms.gov/Newsroom/MediaReleaseDatabase/Press-releases/2016-Press-releases-items/2016-09-08.html

*   **FCC Amateur Radio Service**   http://wireless.fcc.gov/services/index.htm?job=about&id=amateur

*   **HAM radio operators in the US**   http://www.arrl.org/what-is-amateur-radio
❑   Report to US Congress; Uses and capabilities of Amateur Radio Service Communications in emergencies and Disaster relief.
   http://transition.fcc.gov/Daily_Releases/Daily_Business/2012/db0820/DA-12-1342A1.pdf

*   Report & Order US Congress FCC-124
   AMENDMENT OF PART 97 OF THE COMMISSION'S RULES REGARDING   AMATEUR RADIO SERVICE COMMUNICATIONS DURING GOVERNMENT DISASTER DRILLS

http://transition.fcc.gov/Daily_Releases/Daily_Digest/2010/dd100715.html
http://www.calhospitalprepare.org/sites/main/files/file-attachments/amateur   radio order.pdf

HIPAA
Disclosures for Emergency Preparedness - A Decision Tool: Overview
HIPAA https://www.hhs.gov/hipaa/for-professionals/special-topics/
emergency-preparedness/decision-tool-overview/index.html

Editor *– my apologies, the formatting above between different word processors proved difficult. The formatting provided by the author was uniform but when imported here, it became more complicated.*

NOTES

_____

_____

_____

_____

_____

_____

_____

_____

_____

_____

_____

_____

_____

_____

_____

_____

_____

_____

_____

14 AUXCOMM SKILL BRIEFER

Figure – Get folks into a REAL EXERCISE – and learning accelerates!

The usage of the Incident Command System, while widely taught and emphasized throughout the Department of Homeland Security and emergency communications, is not uniform in amateur radio.   ARES groups seem to have wide variability in their familiarity and adherence to this supposedly nationwide system. My experience is relatively "local" – there are other counties where even foot-races are organized with full ICS regalia, but trying to ramp up usage of the ICS system and nomenclature in my county has been a bit of work...

In fact, I've been surprised to find that my local Emergency Operations Center really doesn't carry out full scale exercises!   Yet these are where the "rubber meets the road" for those of us who are NOT full time engaged in law-enforcement, fire-fighting, or disaster first responders.   So lots of opportunities for "interoperability training" never occur.

A key goal of this conference is to break the ice on using the ICS --- to get you familiar with it and to convince you that your local group would be a LOT STRONGER if you began to move into holding some sort of full scale exercise (typically preceded by a training TableTop effort.)

The first problem with the ICS training is that the terms used are new to us, and there are a LOT of the terms, a huge hierarchical organization with "Chiefs" and "leaders" and all kinds of assistants, and Branches and....well it is just kinda confusing.   The treasured training basics of ICS 100, 200, 700 and 800 are filled with a ton of redundancy that doesn't seem to help --- it makes a lot more sense when you finally actually DO the stuff.

   The First Key Part of the ICS system in my view is the ICS201 Incident Briefing.   This key document sets out what the incident is, and what the initial response will be.   Add a bunch more documents to it (ICS 202, 203, 204, 205...) and you have an Incident Action Plan....but you start with the ICS201, so lets look more closely at it:

Blank form for your later use: http://www.qsl.net/nf4rc/BLANK_ICS201_INCIDENTBRIEFING.pdf

Here are the sections:

No.	Section	Explanation
1	Name of the Incident	
2	Number assigned to the incident	
3	Date and time	
4	Map or sketch with much more information	Visually set down the geography of the event.
5	Situation summary and Health and Safety Briefing	Explain what is happening
6	Who prepared this briefing	Person who takes responsibility for this document
7	Current and Planned Objectives	Sets out what we're trying to achieve
8	Current and Planned Actions, Strategies and Tactics	Explains where we're going with this response – items and times!
9	Map of the current organization: Incident Commander Liaison Officer Safety Officer Public Information Officer Planning Section Chief Operations Section Chief Finance/Administration Section Chief Logistics Chief	A basic organizational chart – with standard names and titles. Not all these positions may exist!
10	Resource Summary: a listing of all the resources, whether or when they have or should arrive, and notes on location and assignment.	

Communications comes under the Logistics Chief in the standard organization of ICS.   As you can see, this is a fairly straightforward way to get a good snapshot of what is going on and what the response is. It sets out who is responsible for which areas, and makes assignments of tasks.   It could be used pretty well for a race or marathon.

To make a more complete Incident Action Plan you will want to add some or all of the following documents:

ICS202 – more complete Incident Objectives
ICS203 – more complete Incident Organization
ICS204 – Assignments
ICS205(or a version thereof) Radio Communications (frequencies, modes etc)
ICS206 – Medical plan – extremely useful if you are managing a full scale exercise, because it maps out IN
        ADVANCE where medical help can be found for any of your volunteers as well, level of care, and
        time frames to get care
ICS208 – Safety Plan --- good place to put specifics about the risks of the exercise or eventually

Then there are a few additional "forms" you might want to know about:

ICS 211: a "check-in" form so you know who showed up
ICS 214: Activity Log for documenting what individuals did
ICS 215: Operational Planning
ICS 221 Demobilization Checkoff --- you're not likely to see this except in a large incident, but it has a good
        method for making sure everything got returned, documented, etc.

All of these forms (in one shape or another) are widely available and if you as a leader will begin to use them, you'll understand them far better and your group will be much more accustomed to this organizational system.

Creating an ICS205 on the fly, from an ICS217

CREDIT: This wonderful spreadsheet was freely given out by James Milsap of the Cherokee ARES at the DHS AUXCOM Alabama course :

USAGE:

- **First   populate   the   ICS-217   ("Radio   Frequency   Assignment   Worksheet": https://www.usda.gov/sites/default/files/documents/ics217.pdf) page with ALL your local frequencies.**

- **Then within the ICS-205, merely select the ones that you want to use for your Incident. DONE!**

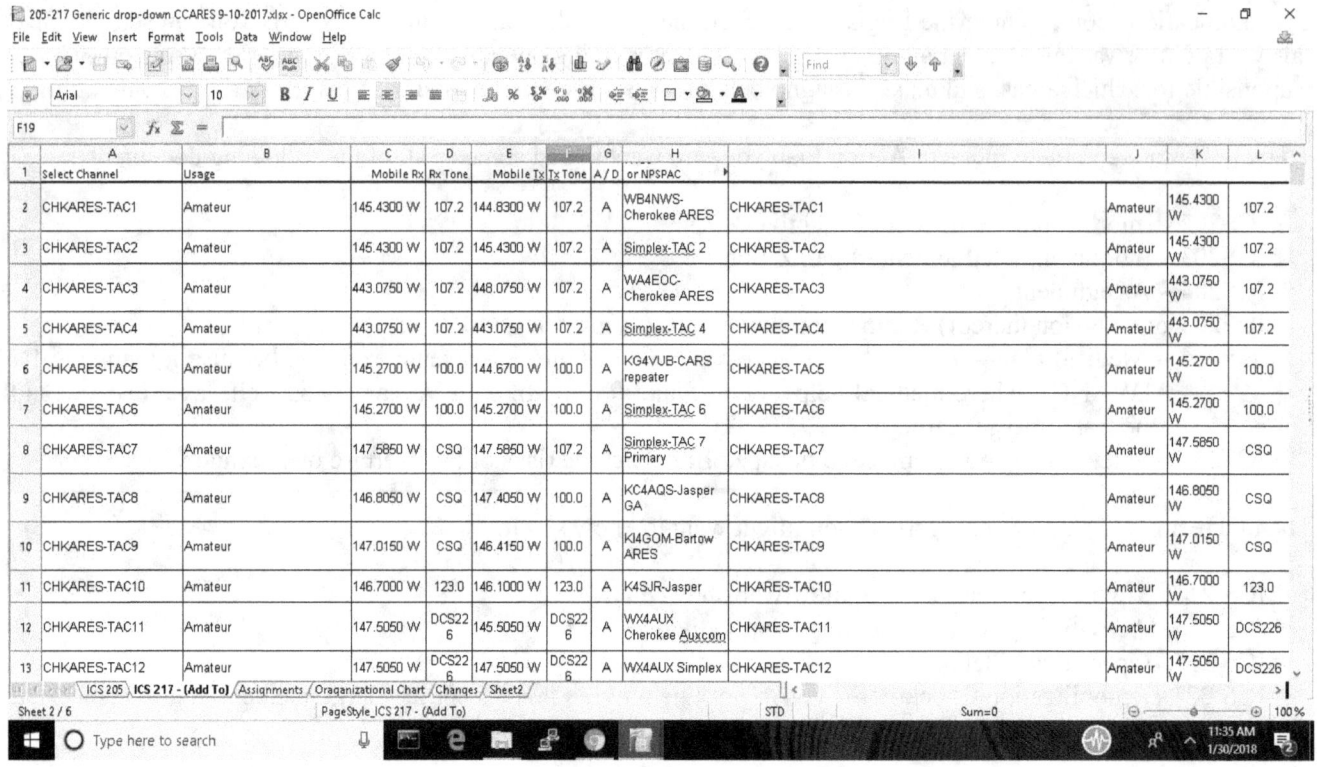

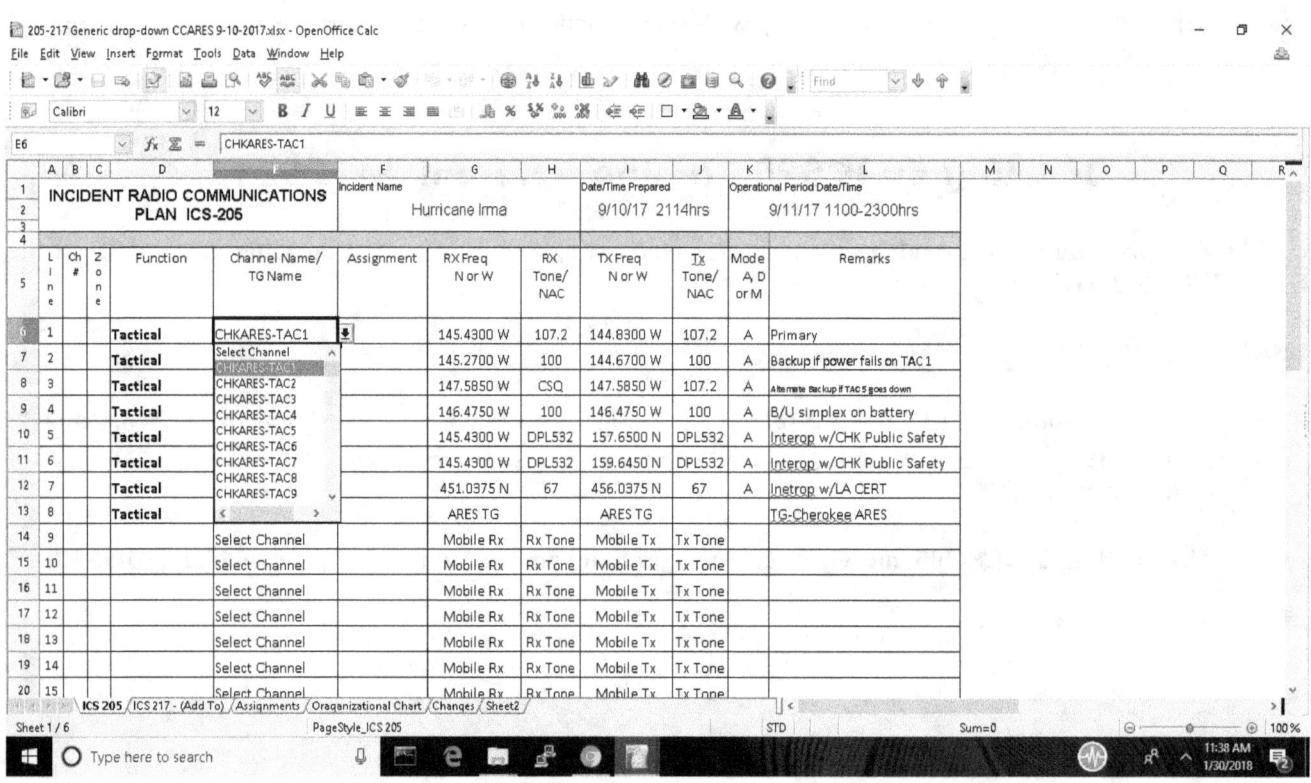

15 EXERCISE PLAN

The goal of this section is to present an HSEEP-type Exercise Plan for our mini-Full Scale Exercise, just to give you a picture of what an Exercise Plan might look like. (See: https://www.fema.gov/media-library/assets/documents/32326)   To save space, different sections do not start on new pages, but just run serially.

2018 Emergency Communications Symposium Mini-Full Scale Exercise
Feb 24 2018
EXERCISE PLAN

EXERCISE OVERVIEW

Exercise Name	(TBA)
Exercise Dates	02/24/18
Scope	This is a Full Scale Exercise carried out as part of the afternoon activities of the 2018 Emergency Communications Symposium. While several Command and General Staff Officers of the Incident Command System will be in place, and as much as possible, ICS procedures will be implemented, this exercise effort primarily involves advanced communications in a disaster scenario.
Mission Area(s)	Response
Core Capabilities	1. Antenna Placement 2. Tactical and Logistics Communications in unusual circumstances 3. Use of variety of modes and bands 4. Backup Power

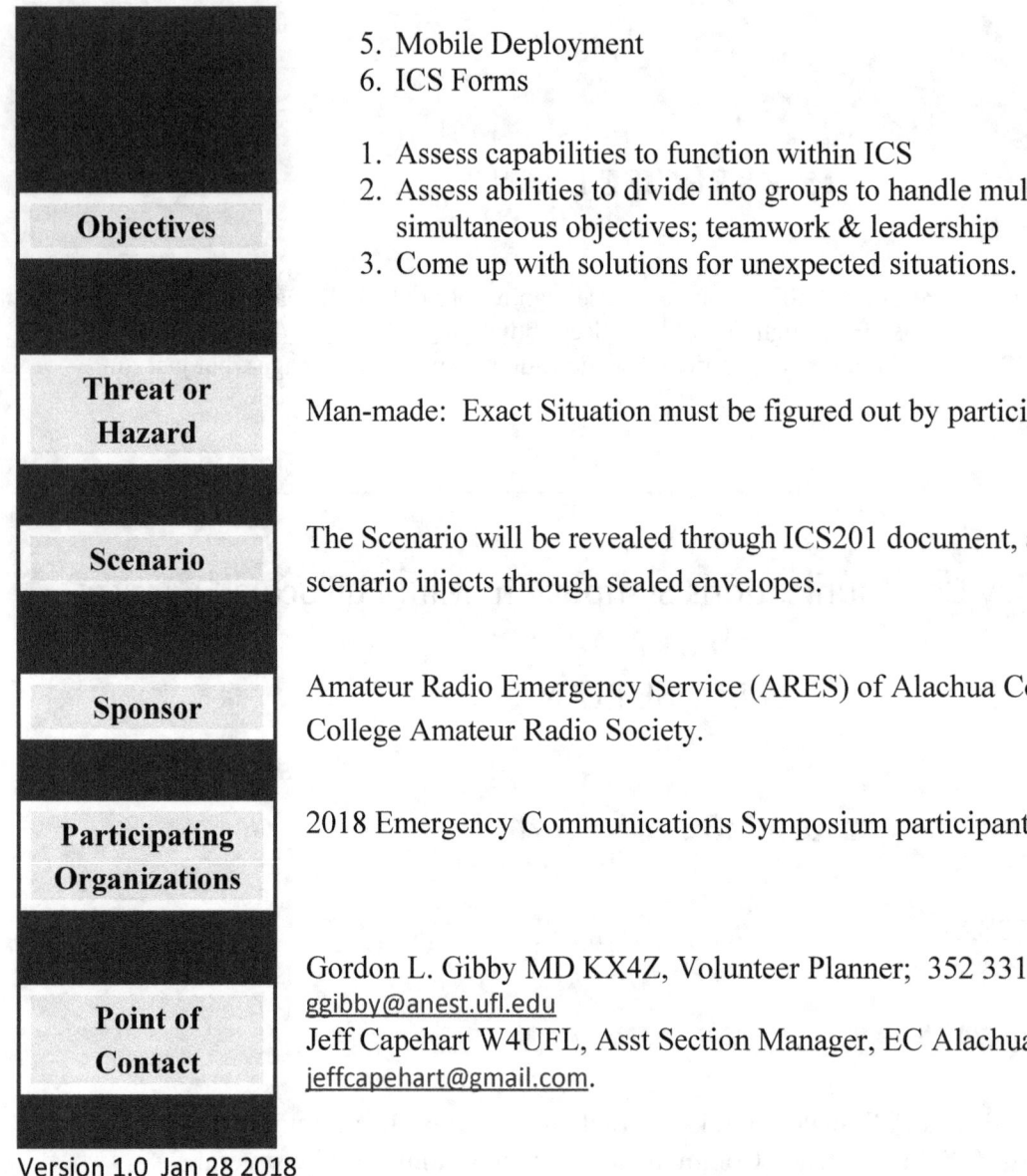

5. Mobile Deployment
6. ICS Forms

1. Assess capabilities to function within ICS
2. Assess abilities to divide into groups to handle multiple simultaneous objectives; teamwork & leadership
3. Come up with solutions for unexpected situations.

Objectives

Threat or Hazard

Man-made: Exact Situation must be figured out by participants.

Scenario

The Scenario will be revealed through ICS201 document, and ongoing scenario injects through sealed envelopes.

Sponsor

Amateur Radio Emergency Service (ARES) of Alachua County / Santa Fe College Amateur Radio Society.

Participating Organizations

2018 Emergency Communications Symposium participants.

Point of Contact

Gordon L. Gibby MD KX4Z, Volunteer Planner; 352 331 6639, ggibby@anest.ufl.edu
Jeff Capehart W4UFL, Asst Section Manager, EC Alachua County, jeffcapehart@gmail.com.

Version 1.0 Jan 28 2018

SECTIONS

Title	Page
Exercise Overview	
Foreword	
Exercise scope, objectives, and core capabilities	
Participant roles and responsibilities	
Rules of conduct	
Safety Issues (including real emergency phrase)	

Logistics	
Security	
Communications	
Schedules	
Maps and Directions	

FOREWORD

Welcome!

Thank you so very much for participating in our training exercise!   Giving of your time and preparation to benefit your community and neighbor are honorable and worthy actions --- and we appreciate your commitment and dedication.

This manual is designed to be read to give you an introduction to our upcoming Exercise – read it first, and then read the ICS-style documents that are being produced for our Exercise as well.

This manual is produced in accordance with the document **Homeland Security Exercise and Evaluation Program (HSEEP), April 2013**, which can be accessed here:

https://www.fema.gov/media-library-data/20130726-1914-25045-8890/hseep_apr13_.pdf

In order to best understand these plans and exercises, it is recommended that the participant take free online courses at the ICS-100, 200, 700 and 800 introductory level.   Addition coursework such as the ICS-300 and ICS-400 are also helpful. There are ICS courses on exercise planning that will help the participant learn more about the development process as well.

Thank you again for your participation!   You're a real asset to your community!

Gordon L. Gibby MD
Newberry, Florida
January, 2018

EXERCISE SCOPE, OBJECTIVES, AND CORE CAPABILITIES

The Goal of Emergency Communications

TO FURNISH EMERGENCY COMMUNICATIONS WHEN REGULAR COMMUNICATIONS FAIL OR ARE INADEQUATE IN THE EVENT OF NATURAL OR MAN-MADE DISASTERS

The Goals of the Full Scale Exercise

PURPOSE
This exercise is designed to provide feedback on our proficiency and capabilities to achieve the likely communications tasks required in a severe communications and mass casualty scenario.

CORE CAPABILITIES

Chosen goal capabilities	Status	Involvement in this exercise
1. Antenna Placement	New group, unknown status.	This exercise will test HF and VHF deployment and possibly some unusual bands.
2. Tactical and Logistical Communications in an unusual environment.	New group, unknown status.	Teams tasked with carrying out multiple types of communications.
3. Use of variety of modes and bands.	New group, unknown status.	Multi-band communications requested.
4. Backup Power	New group, unknown status.	Significant involvement in this exercise
5. Mobile Deployment	New group, unknown status.	Deployment via vehicles is practiced.
6. ICS Forms	New group, unknown status.	Short exercise with only a few ICS forms.

	OBJECTIVES
1	Assess the capabilities of our group to work within the Incident Command System framework.
2	Assess abilities to divide into groups to handle multiple simultaneous objectives; teamwork & leadership
3	Come up with solutions for unexpected problems.

SCOPE --- in multiple aspects

DIMENSION	LIMITATIONS
Kinds of Exercise Participants	Primarily ARES-associated licensed amateur radio operators with prior training but flexible enough to allow untrained amateur radio operators whom we are mentoring to participate; EOC personnel from the Emergency Manager's office where possible.
Geographic Area	Simulated areas within Alachua County, but due to time constraints, using locations on Santa Fe College.
Number of Participants	No more than 50 participants.
Responder Functions	Communication of messages by VHF/HF/other is the bottom line outcome, with process measurement of intermediate functions to achieve the end-goal of communications.
Hazard Type	Unusual man-made communications and multifocal hazard.

PARTICIPANT ROLES AND RESPONSIBILITIES

Participants will integrate into a mythical Incident Command System already in operation to deal with the unusual issues facing them, create subunits and task groups as required.

Our volunteers will staff multiple positions:
- Incident Commander
- Logistics Section Chief
  - Communications Unit #1
  - Communications Unit #2
  - subunits of the Comms Units as deemed appropriate by their Leaders.
  -

Scenario

Similar to how an unexpected emergency may occur, the scenario will be masked until presentation through the ICS201 and ongoing documents. This is unusual for Full Scale Exercises which are more often very carefully explained well in advance.

The group will work out a generic ICS205 in advance, and then be handed the ICS201 of this Exercise and begin immediately.

Exercise Assumptions and Artificialities
Since an exercise isn't reality, there are always assumptions and artificial constraints that are necessary

to make practice possible.   Exercise participants should make reasonable responses to events as they are presented to gain the most training advantage from the exercise.

Assumptions

Some assumptions are required for this exercise:

1.   Participants are here to practice and to learn, mistakes will be made and learned from if we all work together.

2.   Although the events may not be perfectly planned, we accept them as plausible and move forward with the exercise.

3.   If a real world emergency occurs, it certainly takes precedence over our exercise!

Artificialities

During this exercise, the following constraints:

Exercise communication and coordination is limited to participating exercise organizations, venues, and the systems we're trying to test and practice;   refrain from using cell phones!   Further information will be dispensed through three sealed envelopes that are to be opened at defined times throughout the exercise.   The exercise must end on time due to our time constraints, even if the goals have not been fully accomplished.

RULES OF CONDUCT

Participants should observe normal driving regulations and safety precautions at all times. We're on a college campus where firearms are not allowed to be carried except by law enforcement.

SAFETY ISSUES

SHOULD A REAL EMERGENCY OCCUR, USE THE PHRASE "THIS IS A REAL-WORLD EMERGENCY" TO EXPLAIN THAT YOU ARE DEALING WITH AN ACTUAL, RATHER THAN SIMULATED EMERGENCY SITUATION.   THESE COMMUNICATIONS HAVE PRIORITY OVER ALL EXERCISE COMMUNICATION.

TRAVEL: This exercise includes travel by private vehicle.   Check vehicles for fuel, fluids, tires, headlights, brake & turn lights, windshield wipers and other safety equipment prior to departure. Drivers should be aware of the planned route to the Check In point and Staging.   Driver should have a paper map to supplement GPS or other navigational aids, as during real incidents, there may be interference to electronic navigational systems.   Although group communications may be carried on via cell phone or radio, the driver should remain vigilant and undistracted.   Communications duties should be handled by a passenger.

DEPLOYMENT LOCATIONS: Be careful not to damage the deployment locations.   One location may have power tools in place; do not actuate them.   Replace any chairs etc., to their original position before leaving.   Leave everything the way you found it.   Be careful not to create hazards with antennas or transmission lines.

SUN: Susceptible individuals can acquire a painful or dangerous burn in full sun exposure for only minutes.   Protect yourself with awnings, canopies, large-brimmed hats, sunscreen, clothing and wise positioning.

GENERATORS: Check generators for fuel leaks before operation.   Do not refuel a hot generator until it has time to cool somewhat.

ELECTRICAL EQUIPMENT: Be very careful of the currents that low-voltage high-power equipment may require.   All power circuits must have fuses or circuit breakers and correct sizing of wire.   Do not attempt to draw excessive current from a vehicle accessory or cigarette lighter outlet.   Be careful when making connections to vehicle electrical systems to properly connect, avoid reverse polarity, and avoid any kind of dead short across vehicle high current supplies.

ELECTROCUTION: Do not allow water to reach extension cords and connections.

ANTENNA PLACEMENT: Be cautious when emplacing antennas. Be careful if slingshots or other projectiles are utilized.

LOGISTICS

1.   Teams will travel by vehicle to their assigned deployment locations.   Envelope #1 will give further information about this deployment.
2.   When the exercise is over, teams may travel back to the Symposium room by any safe means.

SECURITY

There is no specific "security" for this exercise.   Please keep watch over your communications gear / generators /etc.   Lock your vehicle when you leave it with valuable gear inside of it.   The college campus will likely be deserted but still be aware of your surroundings.

COMMUNICATIONS

1.   Communications will be as per the ICS205 and the scenario constraints.
2.   **SHOULD A REAL EMERGENCY OCCUR, USE THE PHRASE "THIS IS NOT A DRILL; THIS IS A REAL-WORLD EMERGENCY" TO EXPLAIN THAT YOU ARE DEALING WITH AN ACTUAL, RATHER THAN SIMULATED EMERGENCY SITUATION. THESE COMMUNICATIONS HAVE PRIORITY OVER ALL EXERCISE COMMUNICATION.**

SCHEDULES

It is quite possible that the schedule for the day will have been adjusted a bit as the Conference goes on.   At the beginning of the Exercise the closing time will be announced; when that time is reached, the Exercise is over, and you may pack up, clean up, return everything to its rightful place and return to the symposium room by any practicable means.

ADDENDUM: Healthcare Essential Elements of Information (EEI)

- Available means of communications
- Facility operating status
- Staffing status
- Facility structural integrity
- Status of evacuations or sheltering
- Critical medical services (e.g., critical care, trauma)
- Critical service status (e.g., utilities, sanitation, ventilation)
- Critical healthcare delivery status (e.g., bed status, laboratory and radiology)
- Patient/resident transport
- Patient/resident tracking
- Critical/Acute Resource Needs (materials, medications, utility back-up supplies, etc.)

*** *Hospital Operators may be asked to communicate the EEI to another hospital or facility.*

Putting this into table format:

No.	Element	Data
1	Available means of communications	
2	Facility operating status	
3	Staffing Status	
4	Facility structural integrity	
5	Status of evacuations or sheltering	
6	Critical medical services (e.g., critical care, trauma)	
7	Critical service status (e.g., utilities, sanitation, ventilation)	
8	Critical healthcare delivery status (e.g., bed status, laboratory and radiology)	
9	Patient/resident transport	
10	Patient/resident tracking	
11	Critical/Acute Resource Needs (materials, medications, utlility backup-supplies, etc.)	

References:
GA ARES: https://www.gaares.org/documents/HospitalEmergencyOperationsPlan(EOP)-20180114.pdf; also see: https://www.nisconsortium.org/portal/resources/bin/NISC_EEI_Publication_1426695387.pdf

16 FULL SCALE EXERCISES (AND EMP)

A major goal (in my mind at least) of this Symposium is to convince the participants that they really can go back to their local groups and get them doing Full Scale Exercises. I'm just really convinced that THIS is where people really finally get serious about developing the skills, assets and strategies necessary.

- ○ Hurricanes
- ○ Earthquakes
- ○ Wildfires
- ○ EMP
- ○ Other disasters

Jeff Capehart makes the point that it is often the PREPARATION that does the most good: when planning an Exercise for a hurricane, wildfire or something even more terrible (nuclear or EMP attack, biological attack etc) you finally *think through* what might happen ("what if?") and begin to figure out what would be required to deal with it. THAT is where the real preparations often occur.

Hold a TableTop First
Art Grant makes the point that the TABLETOP practices that we hold prior to the Full Scale Exercise (in multiple rooms of one house) allow people to find their errors, get help easily, and correct them.   THAT is where the learning by the participants often occurs.

Take the plunge, decide to do some real exercises in your community, come up with at least an ICS201 / ICS205 about it, hold a table top, then hold the real exercise.   Along the way you'll find it is far easier to ask for help from authorities and other groups – because you have a defined, time-limited goal which just requires a limited commitment, and is obviously for a good cause.

EMP is the elephant in the room.   People are afraid to even think about it.   Throw their hands up in the air! What kind of leadership is that?

Because of the ever-increasing risk from North Korea (among others), the public is becoming more and more educated about the risks of EMP.   QST published all the important information way back in the 1980's only a few years after it was figured out by 1960's actual experiments by the USA and Russians.   (See reprints: http://qsl.net/kx4z/QST-Electromagnetic_Pulse_and_the_Radio_Amateur.pdf)

EMP COMPONENTS

E1	Incredibly strong 50kV/m broadband nanosecond PULSE that destroys semiconductor junctions of anything connected to more than a few feet of "antenna" Bandwidth past 100 MHz. Kilovolts and Kiloamperes for nanoseconds on input stages of radios connected to long feedlines/antennas. Protect for it by hardening – gas discharge tubes, filtering to slow risetime, reduce energy; back to back diodes; vacuum tube gear easy to survive.

	May arc and damage coax. Faraday shield protects equipment but wires penetrating (allowing usage) must be protected. Mil-Stds exist. Military is well protected. More and more gear is protected.
E2	A slower, weaker version of E1. Think more like lightning
E3	Sort of like what Coronal Mass Ejection does. Shoves aside earth geomagnetic field, event lasting minutes. Relative motions of field and long wires produces DC offset currents --- so huge transformers connected to hundred mile long transmission lines get sent into high-loss saturation...catch fire. Grid goes down. Protect by having systems to take you OFFLINE if voltages become whacky (because of loss of neutral wires etc) and have backup power systems not destroyed by E1.

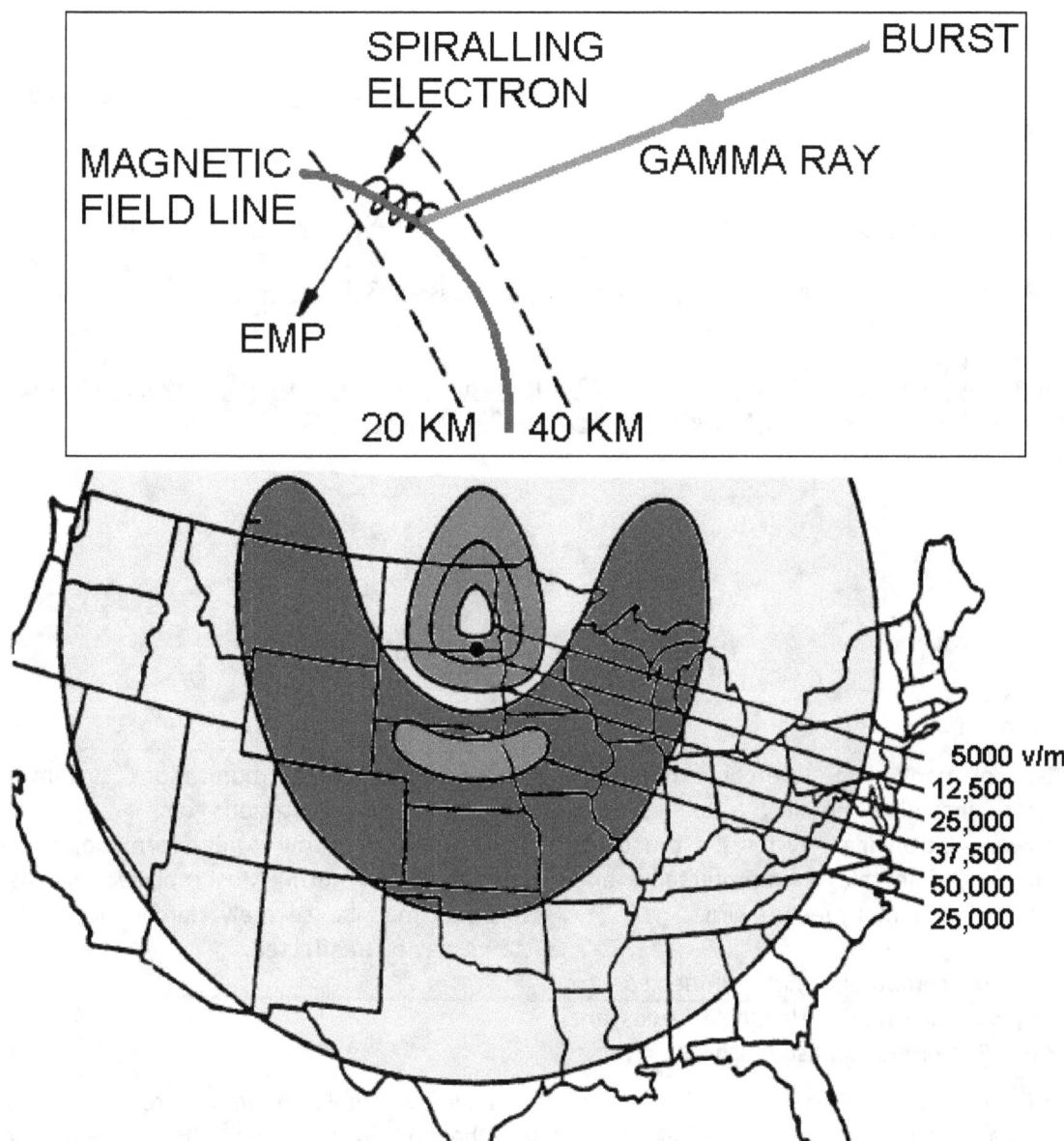

Source: Nuclear Environment Survivability,
U. S. Army, report AD-A278230 (1994)

Mechanism and likely E1 field strength for a 400 km-high EMP nuclear event. Note the extremely strong field over a huge section of the continental USA. Unusual shape is caused by physics of interactions of earth's magnetic field with electrons.

Ref: By User:Photocopier - Wikipedia in english, page "High-altitude nuclear explosion", Public Domain, https://commons.wikimedia.org/w/index.php?curid=37697664

Some really eye-opening expert reports:

2004 EMP Commission report: https://www.globalsecurity.org/wmd/library/congress/2004_r/04-07-22emp.pdf

https://oversight.house.gov/wp-content/uploads/2015/05/Baker-Statement-5-13-EMP.pdf

2008 EMP Commission report: https://fas.org/irp/congress/2008_hr/emp.pdf

http://www.commacademy.org/2015/Session%2020%20-%20EMP%20and%20Other%20Threats%20to%20Electrical%20Infrastructure/Nuclear%20Weapon%20Effects%20to%20Emergency%20Responders%20v3.pdf

Possible major societal problems:

No.	Problem	Response
1	Loss of many solid state communications gear, including possibly satellites, cell towers, trunking systems, broadcast stations, anything unprotected and connected to many feet of wire. Laptop computers not connected to anything are fine. Handheld radios are fine. Repeaters....not so much.	– create alternate communications system to keep public service communications going – major problem is how to get information to the public (rumor and mob control) Must replace broadcasters – Internet may be severely damaged. Fiber optic systems may be unaffected.
2	HF radio communications very limited for hours due to intense D-layer ionization	May have to wait for hours before HF comms will be useful beyond ground wave. VHF connected repeaters might be important---if they survive. Microwave unaffected...if equipment survives.
3	Unpredictable number of vehicles will be partially to fullly disaabled.	Mass traffic accidents, roads, highways and interstates become parking lots making passage exceedingly difficult for remaining working vehicles. Increasingly tiny semiconductor junctions more susceptible; increasingly better ESD protection makes them less susceptible; no one knows exactly what will happen. Older vehicles with fewer electronics may keep running. Diesels, magnetos, points --- much less affected
4	SCADA computer damage leading to catastrophes in piped fluids: natural gas,	Fire response. Alternate water / sewage / fuel solutions

	petroleum, water, sewage	
5	Damage to just-in-time transportation systems for food, fuel, water, products, medical equipment, supplies	Short term: find work arounds. Long term survival depends on finding food sources.
6	Society very vulnerable to attack in any other means due to weakened populace	
7	Chaos, mass riots, lawlessness, starvation.	Congressional committees recommended solutions for most of this. Very little has been done. **Try to keep society stitched together long enough for solutions & help to be found....**

NOTES

17 GROWING YOUR GROUP:
Technician Classes the Immersive Style

Figure: Building antennas teaches all kinds of skills to new prospective hams – and it is CHEAP to do.

by Gordon Gibby KX4Z

ADVERTISE!!
Use every available means to advertise your course. This is a numbers game – must reach N people to get 1 to show up.   Discuss with your friends, fliers on campus, announce at ham radio meetings so others can refer people to the course. Depending on your setting, advertise at local schools, civic organizations, even Public Service Announcements on the radio.

MAKE IT A REAL HAM RADIO INTRODUCTION
To hold an effective Technician License course, I aim not just to get the participants ready to pass the license multi-choice exam --- **but also ready to move right into active ham radio** (particularly emergency communications volunteer service).

In order to accomplish that, I include within the course:

1. experience listening to all kinds of radio communications – voice, cw, digital, FM, SSB, repeater, DX, local
2. as much communicating on the air as possible—hit the local repeater, join the SouthCars net for a few minutes, connect with a local traffic or ragchew net; answer a CQ on voice or PSK or any other mode
3. hands-on experience with simple components (resistors, batteries, LED) and simple VOM
4. hands-on experience with a simple "pine-board" component lab that has two resistors, a capacitor, diode, and LED all connected to a center screw, so you can demonstrate both series and parallel
5. send them home with an inexpensive volt-ohm-meter (Harbor Freight $6)
6. send them home with a pre-programmed Baofeng VHF/UHF transceiver (learn how to program with CHIRP and this is easy to prepare for them)
7. have them MAKE a 2-meter antenna (either a dipole or a Slim Jim)

USE VIDEOS OR MULTIPLE SPEAKERS

Avoid one speaker droning on and on. There are multiple youTube video series online that go over the Technician material. Two that I really like are:

The Ham Whisperer – goes through every question and explains them moderately well
  https://www.youtube.com/user/merryviking

Dave Casler – more "introductory" and doesn't cover everything, intended to go along with a training book, but gives useful demonstrations if you aren't as familiar with an area or don't have the equipment to do an area.
  https://www.youtube.com/channel/UCaBtYooQdmNzq63eID8RaLQ

Art Grant got me into a much better sequence by getting me to do MORE of the Ham Whisperer "question/answer" videos and a bit less of the Dave Casler "sit back and watch" videos --- because now I have enough equipment and experience that I can on the fly show examples of many areas of the radio theory etc.   I like to do one or two of the Q/A videos and then break it up with a 10-15 minute live and hands-on demo. DEFINITELY get them using a volt-ohm-meter!   Here are some demo ideas that work well:

- Explain how the rotary switch on the VOM works, then pass out AA and 9V batteries to measure
- Pass out resistors to measure, or use the pine-board experiment lab. Break them into groups; less threatening for a group to try it.
- Use clip leads and lead them to measure resistors in series, and in parallel
- Have them measure the resistance of a big electrolytic and show how it is charging
- Simple battery – resistor – LED circuit!
- Send morse code with the battery – resistor LED circuit....and then explain it is way up in the GHz and you have a LICENSE to send signals on that frequency!!!
- Antenna impedance measurements with an antenna analyzer --- show resonance
- Tune around bands and let them hear every possible mode
- Send a WINLINK email over HF or VHF
- Open up radios and explain what the various portions accomplish
- Take a zip cord, connect one end to a PL259, connect to antenna analyzer, pull the far ends slowly apart until it resonates on 2 meters --- SWR drops into usable range right in front of their eyes
- Have them slingshot a line over a tree, connect up a random wire antenna and tune it for them to a band using a manual or automatic tuner
- Demonstrate a frequency counter --- there are $10 versions on Ebay
- Build a Slim Jim antenna using #14 house wiring and 1x2 pressure treated wood – always the highlight of our technician courses. Use clips on a special coax jumper to help them find the proper tuning on the stub matching system.

Video list and timings: http://www.qsl.net/nf4rc/CourseMasterVideoTimeDemoList.pdf
Technician Test Questions (with answers)   http://www.qsl.net/nf4rc/2014-2018%20Tech%20Pool.pdf
Baofeng Radio Hints: http://qsl.net/nf4rc/BaoFengHints.pdf

CREATURE COMFORTS
Be sure to have refreshments available.   Make the room temperature comfortable but not sleepy. Have a lunch or other meal prepared.   Salt your meal tables with more ham gear of all kinds.   Use all the time you have to teach.

GETTING PEOPLE TO TAKE THE TEST
If you can run your group through a practice test, you'll significantly increase their chances of actually showing up to take a test. Be certain they know where/when/how to take the test.   Using phone numbers, call and remind them just before the date.   Encourage them to review the questions and even take a practice test --- ARRL and others offer these free.

Above all else, make your course FUN and EXCITING for your students!

The Ham Whisperer List of Question Portions:

<u>Technician Class License Course</u>   (Valid through June 2018)

Lesson 1: T1A Section Amateur Radio services
Lesson 2: T1B Section Authorized frequencies
Lesson 3: T1C Section Operator classes and station call signs
Lesson 4: T1D Section Authorized and prohibited transmissions
Lesson 5: T1E Section Control operator and control types
Lesson 6: T1F Section Station identification and operation standards
Lesson 7: T2A Section Station operation
Lesson 8: T2B Section VHF/UHF operating practices
Lesson 9: T2C Section Public Service
Lesson 10: T3A Section Radio Wave Characteristics
Lesson 11: T3B Section Radio and electromagnetic wave properties
Lesson 12: T3C Section Propagation modes
Lesson 13: T4A Section Station setup
Lesson 14: T4B Section Operating controls
Lesson 15: T5A Section Electrical principles
Lesson 16: T5B Section Math for electronics
Lesson 17: T5C Section Electronic principles
Lesson 18: T5D Section Ohm's Law
Lesson 19: T6A Section Electrical components
Lesson 20: T6B Section Semiconductors
Lesson 21: T6C Section Circuit diagrams
Lesson 22: T6D Section Component functions
Lesson 23: T7A Section Station Radios
Lesson 24: T7B Section Common transmitter and receiver problems
Lesson 25: T7C Section Antenna measurements and troubleshooting
Lesson 26: T7D Section Basic repair and testing
Lesson 27: T8A Section Modulation modes

18 GROWING YOUR EXPERTISE – GENERAL/EXTRA CLASSES

There is SO MUCH to learn about ham radio.   And in order to be "well rounded" and capable for emergency communications, you just can't stop with the Technician license. The General class license opens up effective long-distance HF communications for the emergency communicator.   To really be capable of responding to equipment, antenna, propagation, and emissions difficulties during an emergency, you really need the information of the Extra class license – and then much more!   So don't let the people in your group get stuck at a low level of expertise!

Leadership sets the pace!   If you are a leader, then you need to be "moving up!"   You don't have to know everything right now....but the attitude that you set by your quest for more knowledge, better skills, assets, and strategies....that will be contagious.   Groups grow when there is significance (you're doing something of real value), challenge (you're learning and really doing things) and friendship.

I held a combined General/Extra class course "bootcamp" immersive style over 2 weekends--- 17 hours each. In the end, I think I would have been better to separate General from Extra.   I wasn't able to adequately address each.   However, the knowledge gained by the participants was tremendous and five people went on to get their Extra --- several of whom are leaders in this 2018 Emergency Communications Symposium.   It gave them not only a new grasp of radio, it made them more self-confident!

Some key demonstrations that really help people advance their knowledge:

- **10 meter antenna with insulator taps at center, 33% and 25% – using an antenna analyzer you can show people feedline impedance and how it goes up off center and the antenna becomes usable on other bands.**
- **Antenna analyzer is really crucial. "RF Voltmeter"... and a stock of 25 ohm, 50 ohm, 100 ohm, 200 ohm carbon resistors – start by demonstrating how it can literally measure resistance and move into reactance/impedance and demo with an antenna.**
- **Elegoo electronic breadboard science kits --- allow people to build simple circuits – like a mic preamp (one example: https://www.amazon.com/Project-Basic-Starter-Tutorial-Arduino/dp/B01DGD2GAO)**
- **SEVERAL ham radio stations – many people have never done SSB, or never did CW....or digital.....have hands-on!!**
- **MAKE ANTENNAS --- everyone made their own 40 meter dipole – for real---during our course. We cut the insulators from PVC, drilled the holes, soldered the coax..measured, cut, adjusted...**
- **An oscilloscope or a station monitor scope helps a lot.**
- **Several antennas that allow you to demonstrate resonant antennas, antenna tuners**
- **If possible, some simple transceiver that allows you to talk through how a heterodyne system works**

Note: This is a rough collection of the course, items may not be placed exactly in the right time locations. In general, every 30 minutes or whenever people look sleepy, get them up and have them do something – like walk outside and see antennas, discuss the feedlines, how to assemble. Hams have way too much book learning and *way way too little HANDS on*.   Try to assemble stuff WHILE THEY WATCH – and as soon as they learn something (like crimping) have THEM do it next time.

Sat. AM		Introduction / "Questions" "Geometry" "Losses"
		Look af the Eglin AFB Critical Freq @ 0800
		Small tour of 4 winlink stations upstairs/ explain multiple bands
		Go through General Class Subelement 1 "RULES Questions & Answers
	Casler General 2.1 17'	
	Casler General 2.2 7'	
		Listen to PSK31 on 14 MHz band Demonstrate auto tuner on outside antenna Demonstrate BITX40 transceiver
	Casler Extra 2.1 24'	
	Casler Extra 2.2 13'	
	Casler Extra 2.3 #1 20'	Break somewhere in here During breaks, examine opened equipment
		Hand out voltmeters, explain "danger area" Measure battery voltages on 20VDC Measure 120VAC wall voltage Explain "letting smoke out"
	Casler Extra 2.3 #2 17'	
	Casler Extra 3.1 16'	
		Eglin AFB Critical Frequency check / MUF discuss geometry Measure resistors on voltmeters – series, parallel Transition to "AC OHMETER" – Antenna analyzer – use to measure same series, parallel resistors and note SWR is always a ratio of 50 ohms.
	Casler General 7.1 13'	
	Casler General 7.2 11'	
	Casler General 7.3 17'	
		Somewhere in here do link budgets – 46dBm for 50 W transmitter, -109 receiver from ¼ microvolt sensitivity – discussion of free space loss, > 200 km

	Casler Extra 10.1 8'	
	Casler Extra 10.2 17'	
	Casler Extra 10.3 23'	Now measure SWR of center of test dipole – resonant at 28 MHz, measure R, SWR Move to 1/3 length – remeasure R, SWR – explain results Move to ¼ length – remeasure R, SWR – explain current/voltage distribution along dipole, Z Demonstrate Balun – corrects 1/3 position to 50 ohms SWR 1;1!! Check Eglin AFB Critical Frequency / MUF
	END OF SATURDAY	1
SUN 1 PM		Demonstrate insides of parallel transmission line SWR meter
	Casler General 5.1 Station setup	
	Casler General 5.2	
		Discussion: Direct conversion mechanism / SSB demonstrate crystals on BITX40 Super heterodyne / EMP / lightning protection – pass out Gas Discharge Tubes, explain (5 kAmperes)
	Casler General 5.3 test eqp/monit/2tone	
		Introduction to what blows final amplifiers – lack of impedance matching and high voltages Introduction to controls on Icom radio Team A: bring up 10 meter coax-fed test dipole, make work, call CQ. Team B: advance to same antenna, balanced line fed, manual tuner on 20 meters, call CQ Team A: auto tune on 40 meters with LDB Come up on PSK31 on 20 meters. (this took a lot of work to succeed) Move to outside high antenna, listen to Europe and South America and attempt to work them.
	Casler General 5.4 interference,	
	Casler General 5.4	

	SECOND SATURDAY	
		Explain "transconductance" and how it applies to amplifiers and how the load resistance greatly determines gain. Show basic op amp feed back structure, common emitter gain dependent on ratio of two resistances.
	Casler Extra 4.1 Radio mathematics, introduces quadrature numbers (imaginary)	
		Continue Smith Chart introduction. Demo measuring 25 ohm load at end of 18 foot transmission line --- at various frequencies, reactance / resistance, way more than 100 ohms --- SWR remains constant. Exactly what Smith Chart predicts.
	Casler Extra 4.2 Part 1 19' Electrical Principles	
	Casler Extra 4.2 Part 2 29' Reactance (FINALLY!)	
	Casler Extra 4.2 Part 3 35 ' Resonance	
		Using older tube type radio with LC tuning to peak receiver/transmitter stages, show how amplifier gain depends on plate/collector load.
	Calser Extra 5.1 23' Semiconductor parts	
	Casler Extra 5.2	
	Casler Extra 5.3 21' Digital Logic	
		Demo older TTL logic frequency meter by measuring transmitted 40 meter signal.
	Casler Extra 5.4 14' Optoelectronics	
	Casler Extra 6.1 22' Amplifiers	
	Casler Extra 6.2 33' Signal Processing	
	Casler Extra 6.3 19' Filters & Impedance matching-lite	
		Semiconductor lab: three circuits to build and study 1. various currents through series LED and rectifier diode, draw VI diagram.

		2. LEDs in base and collector leads of common emitter amplifier, visual hfe demo 3. Simple common emitter AC coupled mic preamp with condenser mic
		Begin Antenna building --- teach one person each skill, have them teach others 1. Drilling, sawing insulators from PVC 2. Cutting antenna wire to length 3. Stripping coax, preparing connections 4. Soldering center conductors 5. Tie-ing end insulators
	FINAL SUNDAY	
		With line over tree, and people set up a groups, lay all antennas on ground parallel, hoist one at a time and measure resonant frequency. Fill out paper tag to scotch tape to feedline, indicating actual/desired resonance, % difference, amount of wire to remove (all antennas ended up at lower end of desired band, pretty close to desired) Demo double female barrel connector lengthening of coax lines, and also demo un-un isolator
	Dave Casler Extra Classs antenna videos as much as time remains.	
		Demo prop of 2 meter YAGI and explain how to use in either vertical or horizontal polarization.
	Take both General and Extra Class online randomized test – make up answer sheets to give participants chance to note problem areas.	Later, email them the test they took so they can see the actual questions they had trouble with.

NOTES

19 CREATESPACE PUBLISHING

by Gordon Gibby KX4Z

As your group begins to grow and put together more Full Scale Exercises, you're going to need ways to create inexpensive documents for a larger number of people. Xeroxing works for quite a while – but eventually it gets to be expensive, and binding services at a local copy store tend to be VERY pricey.

When you want to have a report about your group's service or exercise that looks sharp and can be given to city, county or state officials, it is helpful to have a way to bind a document into a book or booklet.

A far cheaper option to create nicely bound ("perfect bound") books at a very inexpensive cost – typically $2 per copy – is to use CreateSpace online publishing, and as a side benefit, you can actually market your group's expertise to help others worldwide. https://www.createspace.com/Login.do

The process of creating a book is really rather simple and mostly involves your creating a document with any word processor (I use free Open Office https://www.openoffice.org/) – then exporting your document to PDF format and uploading into CreateSpace.

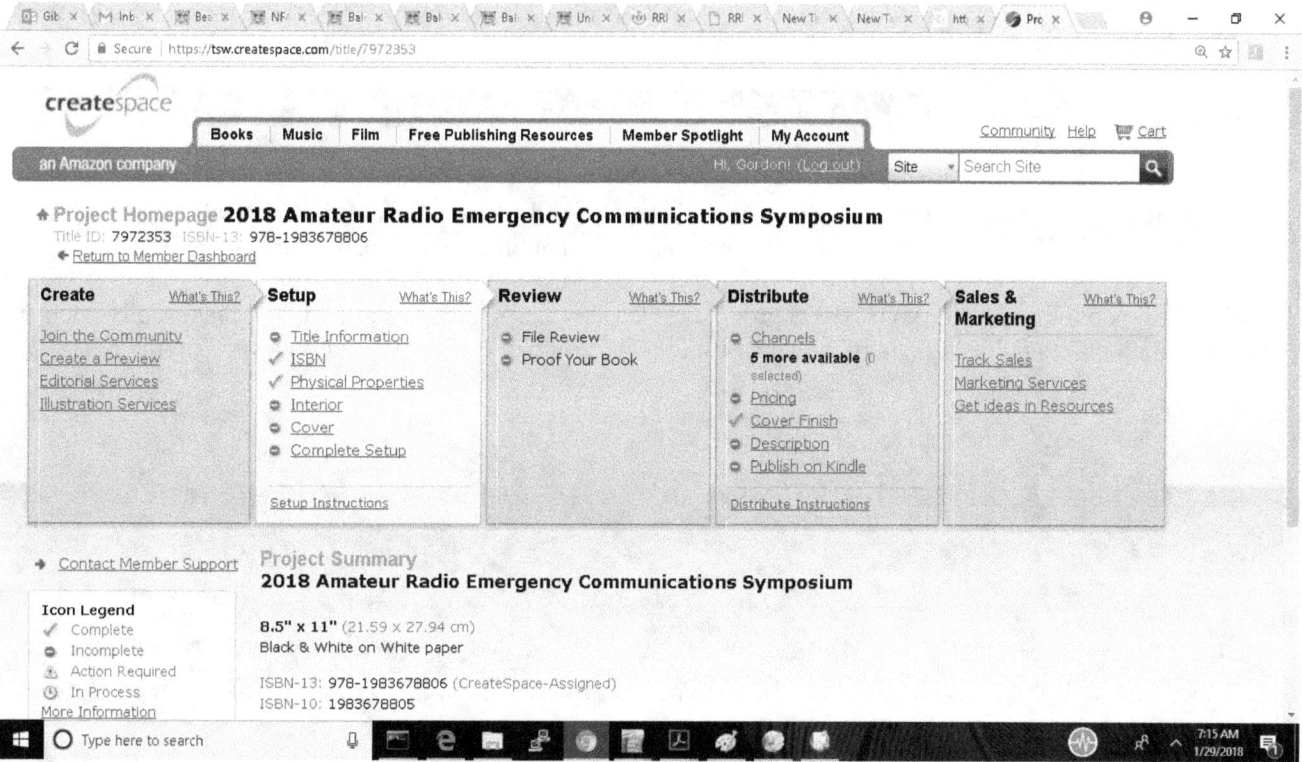

The CreateSpace folks have a fairly straightforward step-by-step process for entering the title, choosing the size of paper, type of paper, black and white (cheap) or color (expensive).   For manuals I use 6"x9" usually, for

reports or larger books (like this one) 8.5 x 11" (normal paper).   They have a cover creator portion that allows
you to upload photos and use stock cover schemes fairly easily.

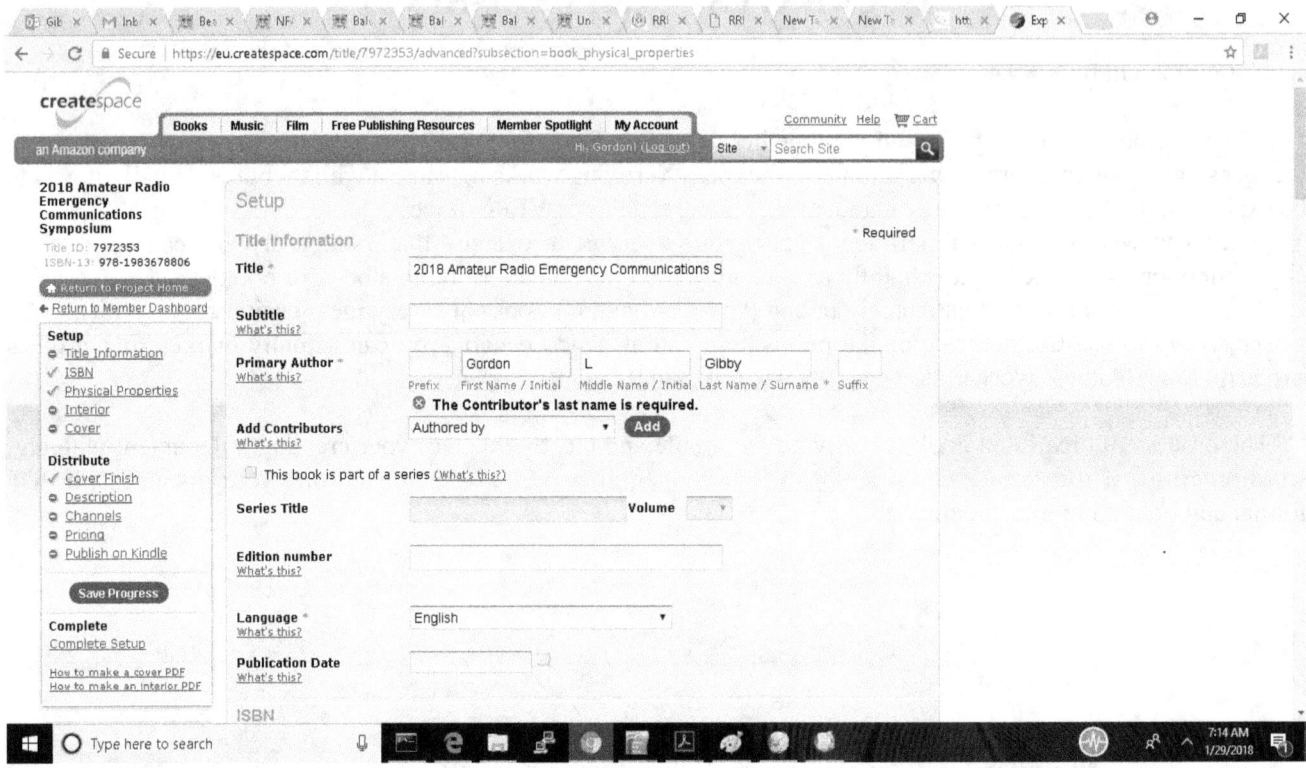

Publishing the book is then little more than choosing a price and where you want it advertised --- everything is
FREE.   It costs NOTHING to publish your text, and there are no publishing companies to veto your work.

Best of all, you the author (or your ARES group!) can order your own copies at cost, and that usually about $2
per copy, which makes publishing training manuals and other group-specific information a very simple process.

ABOUT THE ALACHUA COUNTY
AMATEUR RADIO EMERGENCY SERVICE (ARES) GROUP

The Alachua County ARES group has been having a blast for the last two years!   In the space of about 18 months, we held "antenna parties" and ended up putting up eighteen antennas.   We learned how to solder and build our own sound-card based digital systems and became familiar with AX.25 Packet.   We did WINLINK and also bolstered our voice net capabilities.   About eight or more AX.25 packet nodes sprang into existence, giving us over 4,000 miles of digital coverage, albeit slow.   We connected to – and repaired – a node of a statewide digital net.   We now only have a WINLINK HF RMS server with PACTOR 3, we have three VHF servers and even a Pactor IV SHARES server.   We've done two Full Scale Exercises --- these were not easy! – and held a General/Extra Class course that resulted in five new Extra class licensees. Along the way we had a session on SSTV and we've dabbled with ALE.

Month after month, we just keep learning and trying new things and broadening our experience and capabilities – and we suffer growing pains at times, but we still keep having fun.   And that keeps drawing new quality people to our group, which is dedicated to better and better skills, assets, and strategies for emergency communications. One of my friends spoke of the "conspiracy of competence" – when you start learning how to do things, you draw more people and you gain more allies.

Your group can have all this fun also!

Amateur Radio "Radiogram"

NR	PRECED	HX	Stn of Origin	Check	Place of Origin	Time Filed	Date Filed

Addressed TO: Message Received At:

_____   Station: _____ Phone: _____

_____   Name/Addr: _____

_____   _____

email_____   _____

phone_____   _____

<BT>

____ ____ ____ ____ ____   ____ ____ ____ ____ ____

____ ____ ____ ____ ____   ____ ____ ____ ____ ____

____ ____ ____ ____ ____   ____ ____ ____ ____ ____

____ ____ ____ ____ ____   ____ ____ ____ ____ ____

____ ____ ____ ____ ____   ____ ____ ____ ____ ____

<BT>

SIGNATURE: _____

RCVD FROM		DATE	TIME	SENT TO		DATE	TIME

Amateur Radio "Radiogram"

NR	PRECED	HX	Stn of Origin	Check	Place of Origin	Time Filed	Date Filed

Addressed TO:
_____
_____
_____

email_____
phone_____

\<BT\>

Message Received At:
Station: _____ Phone: _____
Name/Addr: _____
_____
_____

— — — — —   — — — — —
— — — — —   — — — — —
— — — — —   — — — — —
— — — — —   — — — — —
— — — — —   — — — — —

\<BT\>
SIGNATURE: _____

RCVD FROM	DATE	TIME	SENT TO	DATE	TIME

Resources

North Florida Amateur Radio Club (Alachua ARES) web site: http://www.qsl.net/nf4rc/

RRI Traffic Training Manual: http://www.qsl.net/nf4rc/2018/RRI-Digital-Operations.pdf

EMP Reference:
https://www.amazon.com/EMP-Hardened-Radio-Communications-William-Prepperdoc/dp/154077760X

FEMA online training: https://training.fema.gov/is/

ARRL Online Course Catalog: http://www.arrl.org/online-course-catalog

Sarasota Digital Group: http://n4ser.org/sdg/

_____

_____

_____

_____

_____

_____

www.ingramcontent.com/pod-product-compliance
Lightning Source LLC
Chambersburg PA
CBHW081737220526
45468CB00008B/2137